The Humboldtian Tradition

Scientific and Learned Cultures and Their Institutions

Editor

M. Feingold (*California Institute of Technology*)

VOLUME 12

The titles published in this series are listed at *brill.com/slci*

The Humboldtian Tradition

Origins and Legacies

Edited by

Peter Josephson
Thomas Karlsohn
Johan Östling

BRILL

LEIDEN | BOSTON

Cover illustration: Bundesarchiv, Bild 183-36088-0006 / photo: Horst Sturm.

Library of Congress Cataloging-in-Publication Data

The Humboldtian tradition : origins and legacies / edited by Peter Josephson, Thomas Karlsohn, Johan Östling.
pages cm. -- (Scientific and learned cultures and their institutions, ISSN 2352-1325 ; volume 12)
Includes index.
ISBN 978-90-04-27192-0 (hardcover : alk. paper) 1. Universities and colleges--Philosophy. 2. Education, Higher--Aims and objectives. 3. Humboldt, Wilhelm, Freiherr von, 1767-1835--Influence. I. Josephson, Peter, editor of compilation. II. Karlsohn, Thomas, 1967- editor of compilation. III. Östling, Johan, 1978- editor of compilation.

LB2322.2.H87 2014
378.001--dc23

2014011153

This publication has been typeset in the multilingual "Brill" typeface. With over 5,100 characters covering Latin, IPA, Greek, and Cyrillic, this typeface is especially suitable for use in the humanities.
For more information, please see www.brill.com/brill-typeface.

ISSN 2352-1325
ISBN 978-90-04-27192-0 (hardback)
ISBN 978-90-04-27194-4 (e-book)

This book is printed on acid-free paper.

Contents

PART 3
Contemporary Contentions

About the Authors

Mitchell G. Ash

Professor of Modern History. Head of the Working Group in History of Science and Speaker of the multi-disciplinary PhD program 'The Sciences in Historical, Philosophical and Cultural Contexts' (supported by the Austrian Science Fund) at the University of Vienna, Austria. Book publications include: *The Nationalization of Scientific Knowledge in the Habsburg Empire* (1848–1918), ed. with Jan Surman (2012), *Geisteswissenschaften im Nationalsozialismus: Das Beispiel der Universität Wien*, ed. with Wolfram Nieß and Ramon Pils (2010); *Mythos Humboldt: Vergangenheit und Zukunft der deutschen Universitäten*, ed. (1999).

Pieter Dhondt

Senior Lecturer in general history at the University of Eastern Finland. His current research focuses on the history of university celebrations and on medical history, including the process of medicalisation of infant welfare work and the education of health care providers in this field. Among his recent books: *Un double compromis: Enjeux et débats relatifs à l enseignement universitaire en Belgique au XIXe siècle* (2011); *National, Nordic or European? Nineteenth-Century University Jubilees and Nordic Cooperation*, ed. (2011).

Ylva Hasselberg

Professor at the Department of Economic history and director of the Science and Technology Studies Centre at Uppsala University. Recent works include: *Vetenskap som arbete: Normer och arbetsorganisation i den kommodifierade vetenskapen* ('Science as work: Norms and work organisation in a commodified setting') (2012).

Marja Jalava

Academy Research Fellow at the Department of Philosophy, History, Culture and Art Studies, University of Helsinki. Recent works include: 'The Finnish Model of Higher Education Access', in *Fairness in Access to Higher Education in a Global Perspective*, eds. Heinz-Dieter Meyer, Edward P. St. John, Maia Chankseliani and Lina Uribe (2013); 'The Nordic Countries as a Historical and Historiographical Region: Towards a Critical Writing of Translocal History', in *História da Historiografia*, no. 11 (2013); *The University in the Making of the Welfare State* (2012).

Peter Josephson

Researcher at the Department of History of Science and Ideas, Uppsala University. Recent works include: 'Böcker eller universitet? Om ett tema i tysk utbildningspolitisk debatt omkring 1800' ('Books or universities? On a topic in German education policy debate around 1800') (2009), *Universitetet, den lärde och den självlärde* ('Scholars, universities and self-education') (2013) *Historia som kunskapsform – en introduktion* ('Historical knowledge – an introduction') (with Frans Lundgren, 2014).

Thomas Karlsohn

Associate professor, senior lecturer at the Department of History of Science and Ideas, Uppsala University. Recent works include: 'On the Media Conditions of Bildung: Humboldt and Literacy' (forthcoming); *Till vilken nytta? En bok om humanioras möjligheter* ('To what end? A book about the humanities and its possibilities'), ed. with Tomas Forser (2013); *Originalitetens former: Essäer om bildning och universitet* ('Forms of originality: essays on Bildung and university') (2012).

Claudia Lindén

Associate professor, Senior lecturer at the Department of Comparative Literature, Södertörn University. Recent works include 'Virtue as Adventure and Excess. Intertextuality, Masculinity, and Desire in the *Twilight* series', in *Culture unbound: Journal of Current Cultural Research*, vol. 5 (2013); *Att skapa en framtid: Kulturradikalen Anne Charlotte Leffler* ('To Create a Future: The Cultural Radical Anne-Charlotte Leffler'), ed. with David Gedin (2013); 'Är historien alltid redan queer? Om Karen Blixens gotik' ('Is history always already queer? On Karen Blixen's Gothic'), in *Queera läsningar*, ed. Katri Kivilaakso, Ann-Sofie Lönngren, and Rita Paqvalén (2012); 'Temporality and Metaphoricity in Contemporary Swedish Feminist Historiography', in *Rethinking Time: Essays on History, Memory, and Representation*, eds. Hans Ruin and Andrus Ers (2011).

Sharon Rider

Professor, Department of Philosophy and Vice Dean, Faculty of Arts, Uppsala University. Recent publications include: 'In Response to Four Contexts for Philosophy of Education and its Relation to Economic Policy: Three Ideal Types', in *Education, Philosophy and Political Economy,* special issue of *Knowledge Cultures* (forthcoming 2014); 'Philosophy, Globalization and the Future of the University: A Conversation between Sharon Rider and Michael A. Peters', *Review of Contemporary Philosophy*, vol. 12 (2013); *Transformations in Research, Higher Education and the Academic Market: The Breakdown of Scientific Thought*, ed. with Ylva Hasselberg and Alexandra Waluszewski (2012).

Hans Ruin

Professor in Philosophy, Södertörn University (Stockholm). Director of the research program 'Time, Memory and Representation' (www.histcon.se). Recent publications include: *Frihet, ändlighet, historicitet: Essäer om Heideggers filosofi* ('Freedom, finitude, historicity: Essays on Heidegger's Philosophy') (2013); *The Ambiguity of the Sacred: Phenomenology, Politics, Aesthetics*, ed. with Jonna Bornemark (2012); *Fenomenologi, teknik, medialitet* ('Phenomenology, Technology, Medialization'), ed. with Leif Dahlberg (2011); *Rethinking Time: Essays on History, Memory, and Representation*, ed. with Andrus Ers (2011).

Susan Wright

Professor of Educational Anthropology at the Department of Education (DPU), Aarhus University. She studies people's involvement in large scale processes of transformation in universities in Denmark and Britain, and coordinates an EU FP7 ITN project called 'Universities in the Knowledge Economy'. Her latest book is *Policy Worlds: Anthropology and the Anatomy of Contemporary Power*, ed. with Cris Shore and Davide Però (2011).

Johan Östling

Associate Professor of History and Pro Futura Scientia Fellow at the Department of History, Lund University, and Swedish Collegium for Advanced Study (SCAS), Uppsala. Recent publications include: 'The Humboldtian Tradition: The German University Transformed, 1800–1945', in Pieter Dhondt ed., *University Jubilees and University History at the Beginning of the 21st Century* (forthcoming); 'Sweden and the Second World War: Historiography and Interpretation in the Post-War Era', in Jill Stephenson & John Gilmour, *Hitler's Scandinavian Legacy* (2013); 'Humboldts idé: Bildning och universitet i det moderna Tyskland' ('Humboldt's idea: Bildung and the university in modern Germany'), in Jesper Eckhart Larsen & Martin Wiklund, *Humaniora i kunskapssamhället: En nordisk debattbok* (2012).

INTRODUCTION

The Humboldtian Tradition and Its Transformations

Peter Josephson, Thomas Karlsohn and Johan Östling

Humboldt's Ideas

Wilhelm von Humboldt's spirit hovers over the modern university. His name, or at any rate the principles that he is said to represent, is invoked across the world. It is now a little over two hundred years since Humboldt, as a leading Prussian government official, took the initiative in founding Berlin's university, often hailed as the first modern research university. In conjunction with its founding, Humboldt drew up a number of basic principles which many academics still profess to adhere to today; it is most of all in the European context—but also in a global sense—that Humboldt is an obvious reference point, not only in academic speeches and addresses, but also in the wider debate concerning higher education. When the Humboldt-Universität zu Berlin celebrated its centenary in 2010, the legacy of its founder was discussed at numerous conferences and symposia.

What then is Humboldt's legacy? His views on education are often reduced to a series of fundamental principles, which describe how a university should be organized and what its general aims should be. These are most clearly set out in his memorandum, 'Über die innere und äussere Organisation der höheren wissenschaftlichen Anstalten in Berlin'.[1] A short document, barely ten pages when printed, it has been called 'perhaps the most discussed document in the modern history of universities'.[2] One of the crucial ideas it conveys is Humboldt's belief that the inner life of any seat of learning must be one of freedom of thought and education. Students should be free to study under any

1 Wilhelm von Humboldt, 'Über die Innere und äussere Organisation der höheren wissenschaftlichen Anstalten in Berlin', in id., *Werke in fünf Bänden*, iv (Darmstadt: Wissenschaftliche Buchgesellschaft, 2010). Edward Shils has made an English translation of the text: Wilhelm von Humboldt, 'On the Spirit and Organisational Framework of Intellectual Institutions in Berlin', *Minerva* vol. 8 (1970).

2 Björn Wittrock, 'The Modern University: Three Transformations', in Sheldon Rothblatt & Björn Wittrock (eds.), *The European and American University since 1800* (Cambridge: CUP, 1993), 317 n. 33.

 | DOI 10.1163/9789004271944_002

teacher they wanted: higher education, according to him, was radically different from the school system, and when one entered the portals of a university one was an adult, a legally competent individual, with both the freedom and the responsibility to direct one's own studies. Professors, in turn, should have the right to lecture on any subject and to promote new truths, even if these were uncomfortable for the political and religious authorities. In Humboldt's model, the position of professor was considerably strengthened, even though the state still retained the right to determine appointments.

It was also a central idea of Humboldt's that the production of knowledge is a collective and joint concern, and one that should take place at a certain distance from society. One oft-quoted formulation in his university memorandum runs that knowledge should be produced in 'isolation and freedom' (*Einsamkeit und Freiheit*), and with the pursuit of pure knowledge as the guiding ideal.[3] The university was to be a secluded place, which, precisely because of its distance from the surrounding world, would be able to make a significant contribution—to be of benefit to society. For Humboldt, the successful development of knowledge was dependent on a university having a clear boundary separating it from its surroundings. At the same time, a sense of community was important to Humboldt, who regarded research and teaching as mutual prerequisites. At a university, a teacher could be part of a social and pedagogical context, and thus avoid the dangers of isolation. New knowledge could be tested and developed together with the students. At the same time, the students would receive a better education because their lecturers would be able to follow and understand the latest developments in their fields. The search for knowledge should be a dialogic, organic process in which all participants are dedicated to seeking the truth.

Of all his concepts, this is best captured in Humboldt's idea of combined education and self-development, *Bildung*.[4] This is a term that attempts to encapsulate both the individual's personal incorporation of knowledge in their

---

3 Humboldt 2010, 255.

4 A fair picture of his idea of *Bildung* is given in Wilhelm von Humboldt, 'Theorie der Bildung des Menschen: Bruchstück', in id., *Werke in fünf Bänden* (Darmstadt: Wissenschaftliche Buchgesellschaft, 2010), i. 234ff; see also Dietrich Benner, *Wilhelm von Humboldts Bildungstheorie: Eine problemgeschichtliche Studie zum Begründungszusammenhang neuzeitlicher Bildungsreform* (Munich: Beltz Juventa 2003); David Sorkin, 'Wilhelm von Humboldt: The Theory and Practice of Self-Formation (*Bildung*) 1791–1810', *Journal of the History of Ideas*, 44/1 (1983); on the meaning of *Bildung* to Humboldt's contemporaries, see, for example, Rudolf Vierhaus, 'Bildung', in Otto Brunner, Werner Conze & Reinhart Koselleck (eds.), *Geschichtliche Grundbegriffe: Historisches Lexikon zur politisch-sozialen Sprache in Deutschland* (Stuttgart: Klett-Cotta, 1972), i. 508–551.

own unique existence, and the university's joint efforts in the development of knowledge. It also embodies the insight that knowledge can never be complete, and that learning is an open-ended process. In this way, Humboldt's conception was clearly different from earlier thinking, which had regarded knowledge as a fixed, finite body of material that it was the university's task to convey.

When Humboldt's name comes up in modern discussions, it is often in reference to these rules of thumb for academic life; however, it is important to clarify exactly which Humboldt people are talking about when they look to him to support their arguments in contemporary issues. In our experience, few of the politicians and academics who claim to nurture his legacy today take the trouble to distinguish between the historical person and the revered, iconic symbol of the same name.

In this book, we seek to enliven the discussion of Humboldt the symbol *and* the historical character. We thus draw on the growing body of literature that in recent years has problematized traditional attitudes towards both the development of the modern research university and its putative founder. What until quite recently was thought to be a readily identifiable tradition, unequivocally derived from Humboldt and his work, is today a considerably more multilayered phenomenon. At the same time, we also seek to combine the latest research with the educational and research policies of our own day. The positive effect of the historical legacy is no longer as self-evident as it once was in speechifying and research policy debate. The question urgently in need of an answer is how we now ought to relate to the ideas associated with Humboldt's name. What is Humboldt's relevance to the twenty-first century?

Humboldt and the Darkness of the Eighteenth Century

Let us begin in the past; specifically, in the eighteenth century. There is a general consensus among historians today that before Humboldt the German university was an institution in eclipse. Much is often made of the fact that at the turn of the nineteenth century, teaching was conducted in the same manner as it had been in the Middle Ages: the lecturer lectured, the students took notes in silence, the mute recipients of an encyclopaedic educational package, an unchanging repertoire of received Aristotelian doctrine, handed down to them from the lectern. Family ties and patronage determined who could become a professor. The students' greatest pleasure, beside drink, gambling, and prostitutes, was the harassment of the local populace. The usual picture of life at eighteenth-century universities is summarized by Thomas Nipperdey in the first part of his *Deutsche Geschichte* from 1983:

> The universities in the late eighteenth century were generally ossified and decayed, locked into a corporate guild mentality, group privilege, and patronage; study and teaching were frequently scholastic, pedantic, at best encyclopaedic; the students, many of whom were unsuitable, often lived in a brutish, 'young roarer' subculture and moral decrepitude, which found its outlet in rioting and terrorizing the townspeople.[5]

Nipperdey and other historians freely admit that things were not equally bad everywhere, and are at pains to stress that from an early stage the universities of Halle and Göttingen were exceptions to the rule. There, teachers were expected not only to instil in students a pre-approved curriculum, but also to drive knowledge forward. From the turn of the eighteenth century on, Jena too distinguished itself as a vital academic environment. The significance of these exceptions is often played down, however. 'For most contemporaries', explains Thomas Albert Howard in a study from 2006, 'the seeds of change were obscured by the reality of stagnation. Indeed, the century preceding the founding of the university of Berlin was no golden age for German universities'.[6]

Against this background, Humboldt is thought of striding into the picture as a saviour. The historical depictions have more than a little of the heroic epic about them: all was darkness—and then came Humboldt. Whether or not this fable is supported by the evidence, the picture of the eighteenth century as a gloomy epoch in the history of the universities is essential in achieving the epic effect. Without a backdrop of darkness, the hero would not be thrown into the proper relief in all his glory.

Yet it should be borne in mind that in the history of the universities the eighteenth century has not always been painted in such dismal tones. In his books published at the turn of the last century, Friedrich Paulsen—long hailed as the leading authority in the field—presented a very different picture to the one common today. What most of all distinguishes Paulsen from many of his later colleagues is that he depicted the Enlightenment universities of Halle and Göttingen not as isolated points of light in a sea of darkness, but as two model institutions that rulers and governments across the whole of German-speaking

5 Thomas Nipperdey, *Deutsche Geschichte 1800–1866: Bürgerwelt und starker Staat* (Munich: C. H. Beck, 1994), 57.

6 Thomas Albert Howard, *Protestant Theology and the Making of the Modern German University* (Oxford: OUP, 2006), 80f.

Europe would soon wish to copy. 'At the end of the eighteenth century the entire German university world was reformed in accordance with the Halle and Göttingen models', as he wrote.[7] Teaching was no longer conducted in Latin, but in German; extemporary lectures and seminars replaced reading aloud from well-thumbed tomes; teachers were expected to conduct research as well as to teach; governments increasingly respected academic freedom; philosophy, not theology, was in many places regarded the leading form of knowledge. Paulsen concludes:

> Thus suffused with the modern spirit, the German universities took up the position in our public and intellectual life that they have occupied ever since. As the French shrugged off their old universities in the Revolution once and for all, and as in England, where obsolete, old-fashioned institutes were looked down upon as behind the times, hardly more fitting than the schoolbook drills of youth, so at the same time the German people looked hopefully to their universities, looking to them to set the pattern in matters of knowledge and *Weltanschauung*, and in the very functioning of public life.[8]

In short, in stark contrast with the descriptions common in more recent historical writing, Paulsen depicts the eighteenth-century German university as an essentially dynamic and vital institution. True, Humboldt has a place in Paulsen's account, but his contribution appears less exceptional than we are used to today. It is the sections on the eighteenth century—not those on the nineteenth—which fall under the heading 'Die Entstehung der modernen Universität', the rise of the modern university.

It is debateable which of these two descriptions is nearest the truth. The question is not easily settled. During the second half of the eighteenth century, the university was a controversial institution. A small but influential faction, with, among others, the Humboldt family's former private tutor Joachim Heinrich Campe in the vanguard, painted the university as a moral quagmire, and described the teaching on offer there as old-fashioned and detached from the world. Another body of opinion, with adherents mostly among academics themselves, described the same institutions as intellectually dynamic and generally indispensable, not just for the ruling powers, but also for knowledge itself. There is a substantial difference between how university affairs are

---

7 Friedrich Paulsen, *Das deutsche Bildungswesen in seiner geschichtlichen Entwicklung* (Leipzig: B. G. Taubner, 1906), 76.

8 Ibid. 77.

described in, for example, the Göttingen philosopher Christoph Meiners's *Schutzschrift für den Stand und die Lebensart der Professoren* from 1776 and the reformist educationist Christian Gotthilf Salzmann's critical novel *Carl von Carlsberg oder das menschliche Elend,* which was published in six volumes between 1783 and 1787. Regardless of which version of history has later been preferred, it has always been possible to find support for it in the sources.

One thing is evident, however: Paulsen's account of developments in the eighteenth century is harder to square with the image of Humboldt as the creator of the modern university. Similarly, it therefore appears to be more than a coincidence that this version of history was quickly phased out and replaced with depictions of the kind represented here by Nipperdey and Howard. When Paulsen was writing, at the turn of the twentieth century, Humboldt's name was rarely mentioned in discussions of higher education; as Sylvia Paletschek has shown, his posthumous career as an academic founding father first gained momentum because of the celebration of the centenary of Berlin's university in 1910.[9] At the same time as Humboldt's star began to rise, that of the eighteenth century fell. The reassessment of the eighteenth-century universities at the start of the twentieth century was not so much a result of refined source criticism and careful research as the expression of a hagiographical longing.

In recent years, the balance of light and shadow in this portrait has once again begun to shift. Many historians, not only Sylvia Paletschek, have explained that the image of Humboldt as the pre-eminent university reformer is an invention of relatively recent date, and that it does not really survive close scrutiny.[10] It is easy to see that such a critical re-evaluation must also have consequences for how we interpret the age that preceded Humboldt, and that in turn there will be further consequences for how we judge Humboldt's importance. The one influences the other. A study of the history of universities that has rightly attracted a great deal of attention in recent years is William Clark's *Academic Charisma and the Origins of the Research University* of 2006.[11] Perhaps the most remarkable thing about Clark's investigation, at least in the present

9 Syliva Paletschek, 'Die Erfindung der Humboldtschen Universität: Die konstruktion der deutschen Universitätsidee in der ersten Hälfte des 20. Jahrhunderts', in *Historische Anthropologie* 10 (2002), 183–205; Sylvia Paletschek, 'Verbreitete sich ein "Humbold'sches Modell" an den deutschen Universitäten im 19. Jahrhundert?', in Reiner Christoph Schwinges (ed.), *Humboldt international: Der Export des deutschen Universitätsmodells im 19. und 20. Jahrhundert* (Basel: Schwabe, 2001b), 75–104.

10 See note 22 below.

11 William Clark, *Academic Charisma and the Origins of the Research University* (Chicago: University of Chicago Press, 2006).

context, is that it depicts the growth and development of the modern research university in Germany without Humboldt appearing in anything more than a minor role. Instead, the process is shown as being driven by rulers and government officials who had been schooled in the principles of cameralism, and who during the eighteenth century deliberately reformed the archaic routines and practices that then characterized the academic world.

The Turn of the Nineteenth Century

However one evaluates Humboldt's contribution or his importance in relation to the age which preceded the foundation of the university in Berlin, there is every reason to take note of the historical context in which he found himself, and his ambitions when he took over responsibility for Prussian education. The immediate background to Humboldt's establishment of the new university was the desperate condition in which Prussia languished after the twin battles of Jena and Auerstedt in October 1806. The Prussian army had been virtually annihilated, and Napoleon had subsequently taken Berlin. King Friedrich Wilhelm III was forced to flee to Königsberg in what was then East Prussia, and, at the same time, an intensive process of reform was started, affecting many aspects of society, education policy among them. It was in these circumstances in early 1809 that Humboldt unwillingly agreed to take up the leadership of the government department for education and culture. By that point, the issue of founding a university in Berlin had been debated back and forth for years. Now there was an opportunity to turn words into action, and establish a university in a Berlin environment that historians have described as 'a wasteland, scientifically speaking'.[12]

When Humboldt took up his post in Berlin, he had already definite ideas on what an ideal university could and should be. If nothing else, the time he spent in Jena in the 1790s had played a role in forming his views on higher education:[13] a Jena that was perhaps the pre-eminent intellectual powerhouse of German-speaking Europe, and where the Romantic concepts of the sociality of

12 Walter Rüegg, 'Themes', in id. (ed.), *A History of the University in Europe* (Cambridge: CUP, 2004), iii. 16.

13 For Humboldt in Jena, see Helmut G. Walther, 'Reform vor der Reform: Die Erfahrungen Wilhelm von Humboldts in Jena 1794–1796', in Rainer Christoph Schwinges (ed.), *Humboldt International: Der Export des deutschen Univärsitetsmodells im 19. und 20. Jahrhundert* (Basel: Schwabe, 2001); and, for example, Herbert Scurla, *Wilhelm von Humboldt: Werden und Wirken* (Düsseldorf: Claassen, 1976), 134ff and 166ff.

knowledge and of the importance of multipersonal dialogue for education and culture had developed. These ideas, although not speaking to the condition of the universities, or even formulated there, could nevertheless be transferred to higher education.

In the Jena of the turn of the nineteenth century, many of the lectures were held which with time would come to be seen as central to the intellectual history of the university, and which in many respects were mirrored by Humboldt.[14] Schiller was first, with his *Was heißt und zu welchem Ende studiert man Universalgeschichte?* (1789), in which he condemned what he called *Brotgelehrte* ('bred-fed scholars') and their rote learning, while at the same time honouring the philosophically inclined man.[15] A similar tone may be found in Fichte's lecture series *Einige Vorlesungen ueber die Bestimmung des Gelehrten*, held five years later (1794), and, not least, in the young Schelling's *Studium Generale* of 1803.[16] In this last series of lectures, Schelling clearly articulates the connection between research and teaching. He likewise advocated extensive academic freedoms, which required that the state keep away from the internal matters of the university. The programme which Humboldt drew up when the establishment of the university in Berlin had thus to a large degree been anticipated by his predecessors in Jena. Theodore Ziolkowski even writes of an 'institutionalization of the Jena Ideal' in the Prussian capital.[17] Humboldt, however, also drew influence from the writings on university policy produced in Berlin in the years before the founding of the university. One example is Fichte's *Deduzierter Plan einer zu Berlin zu errichtenden höheren Lehranstalt* (1807); another is Friedrich Schleiermacher's *Gelegentliche Gedanken über Universitäten in deutschem Sinn* (1808).[18]

---

14 See Michael J. Hofstetter, *The Romantic Idea of a University: England and Germany, 1770–1850* (Basingstoke: Palgrave, 2001), 22ff; Theodore Ziolkowski, *German Romanticism and Its Institutions* (Princeton: Princeton Unversity Press, 1990), 237ff.

15 Friedrich Schiller, *Was heißt und zu welchem Ende studiert man Universalgeschichte?* [1789], in *Schillers Werke: Nationalausgabe*, ed. Karl-Heinz Hahn, xvii (Weimar: Hermann Böhlaus Nachfolger, 1970).

16 Johann Gottlieb Fichte, *Einige Vorlesungen Über die Bestimmung des Gelehrten* [1794], in id., *J. G. Fichte Gesamtausgabe*, ed. Reinhard Lauth, Hans Jacob & Richard Schottky, i/3 (Stuttgart-Bad Cannstatt: Friedrich Frommann, 1966); Friedrich Joseph Wilhelm Schelling, *Studium Generale* (Stuttgart: Kröner, 1954).

17 Ziolkowski 1990, 286.

18 Johann Gottlieb Fichte, 'Deduzierter Plan einer zu Berlin zu errichtenden höheren Lehranstalt' [1807], in Ernst Anrich (ed.), *Die Idee der deutschen Universität. Die fünf Grundschriften aus der Zeit ihrer Neubegründung durch klassischen Idealismus und romantischen Realismus* (Darmstadt: Hermann Gentner, 1956), 127–217; Friedrich Schleiermacher,

Thus, when Humboldt arrived in Berlin at the end of February 1809, the ground had long been prepared, and the central elements in the system of thought he would soon launch had already been formulated by others. Yet, unlike his fellow enthusiasts, Humboldt was of noble rank; he was able to devote himself to serving the state by trying to put the ideas of the university into tangible effect. Thus, only a few months after his arrival, he was in a position to petition for the foundation of a new seat of learning. In his request, he emphasized the significance to Prussia of being able to establish a new haven for knowledge at a time of national defeat.[19] On 16 August 1809, the king signed the document that founded the new university.[20]

It took slightly over a year before teaching and research could start at *Alma Mater Berolinensis*. The venture started on a small scale in October 1810, and it was only after six years that any actual statutes were drawn up. Soon, however, the new university won academic renown, mainly thanks to Humboldt's success in recruiting acclaimed scholars to take up many of the important chairs. Fichte became the first holder of the crucial chair in philosophy (and in addition was the university's rector for a short time), and he was succeeded in 1818 by Hegel.

By that time, Humboldt had long since left the Prussian educational system behind him: instead, he had become a diplomat, but by the end of the 1810s he withdrew from public life, settled in Tegel, and dedicated much of the remainder of his life to extensive linguistic studies.[21]

'Gelegentliche Gedanken über Universitäten in deutschem Sinn: Nebst einem Anhang über eine neu zu errichtende' [1808], in Ernst Anrich (ed.), *Die Idee der deutschen Universität: Die fünf Grundschriften aus der Zeit ihrer Neubegründung durch klassischen Idealismus und romantischen Realismus* (Darmstadt: Hermann Gentner, 1956).

19 Wilhelm von Humboldt, 'Antrag auf Errichtung der Universität Berlin', in id., *Werke in fünf Bänden* (Darmstadt: Wissenschaftliche Buchgesellschaft, 1982), iv. 113ff.

20 A new scholarly history of the University of Berlin has recently been published. For a detailed account of the foundation of the university, see Rüdiger vom Bruch & Heinz-Elmar Tenorth (eds.), vol. 1: *Gründung und Blütezeit der Universität zu Berlin 1810–1918* (Berlin: Akademie Verlag, 2012). The other volumes include: Heinz-Elmar Tenorth & Michael Grüttner (eds.), vol. 2: *Die Berliner Universität zwischen den Weltkriegen 1918–1945* (2012); Konrad H. Jarausch, Matthias Middell & Annette Vogt (eds.), vol. 3: *Sozialistisches Experiment und Erneuerung in der Demokratie – die Humboldt-Universität zu Berlin 1945–2010* (2012); Heinz-Elmar Tenorth (ed.), vol. 4: *Genese der Disziplinen: Die Konstitution der Universität* (2010); Heinz-Elmar Tenorth (ed.), vol. 5: *Transformation der Wissensordnung* (2010); Heinz-Elmar Tenorth (ed.), vol. 6: *Selbstbehauptung einer Vision* (2010).

21 Paul R. Sweet, *Wilhelm von Humboldt: A Biography*, 2 vols. (Columbus, Ohio: Ohio State University Press, 1978–1980); Scurla 1976.

Humboldt's Nineteenth Century

When Wilhelm von Humboldt died on 8 April 1835, he was hailed as a statesman, a writer, and an educational reformer. Yet the fact is that he never became a point of reference in nineteenth-century German discussions on the university. His fame came later. This is a crucial insight of the new research into Humboldt's legacy conducted by historians such as Mitchell G. Ash, Rüdiger vom Bruch, Walter Rüegg, and Sylvia Paletschek over the last two decades.[22] The one who has consistently argued that Humboldt was absent from the nineteenth century is Paletschek: she has written of 'the invention of Humboldt', backed up by her thoroughgoing investigations of Tübingen University under the German Empire and the Weimar Republic, and maintains that Humboldt's programmatic texts remained unknown or were simply left unpublished. Accounts of university history invoked the writings of Schleiermacher, Fichte, and Steffens, which had been published at the time Berlin's university was being founded. Rarely, if ever, was Wilhelm von Humboldt's name mentioned.[23]

The university in Berlin, which in 1828 was renamed the Friedrich-Wilhelms-Universität, was also in many ways just one university among many. Nothing in its statutes suggested that a new kind of university had seen the light of day. Despite the fact that many had been eager to see a different kind of organization, the faculty hierarchy was the same as before—theology, law, medicine, and, crucially, philosophy—neither was Berlin noticeably different from other German institutions when it came to administrative structure, examinations, and the professors' specializations.[24] Similarly, the university in Berlin was hardly an exemplary model in the intellectual discussions of the nineteenth

22 Walter Rüegg, 'Der Mythos der Humboldtschen Universität', in Mathias Krieg & Martin Rose (eds.) *Universitas in theologia–theologia in universitate: Festschrift für Hans Heinrich Schmid*, (Zürich: Theologischer Verlag, 1997); Rüdiger vom Bruch, 'A Slow Farewell to Humboldt? Stages in the History of German Universities, 1810–1945', in Mitchell G. Ash (ed.), *German Universities Past and Future: Crisis or Renewal?* (Providence, RI: Berghahn, 1997); Paletschek, 2001b; Sylvia Paletschek, *Die permanente Erfindung einer Tradition: Die Universität Tübingen im Kaiserreich und in der Weimarer Republik* (Stuttgart: F. Steiner, 2001a); Sylvia Paletschek, 'The Invention of Humboldt and the Impact of National Socialism: The German University Idea in the First Half of the Twentieth Century', in Margit Szöllösi-Janze (ed.), *Science in the Third Reich* (Oxford: Berg, 2001c); Marc Schalenberg, *Humboldt auf Reisen? Die Rezeption des 'deutschen Universitätsmodells' in den französischen und britischen Reformdiskursen (1810–1870)* (Basel: Schwabe, 2002).

23 Paletschek 2001b, 94ff.

24 Paletschek 2001b, 79f.

century. In the handbooks, encyclopaedia, and compendia of the day, neo-Humanist ideas and Prussian university reforms were not depicted as turning-points. On the contrary, the birth of the modern university was credited to the Enlightenment rationalism of Göttingen and Halle.[25]

The debate on Germany's universities continued throughout the nineteenth century. There was a general consensus that the qualities that above all others set the German universities apart were academic freedom and teaching based on theoretically informed scholarship. When it was necessary to define the purpose of a university, the emphasis consistently fell on the transmission of knowledge and the training of lawyers, theologians, and doctors. It was only towards the end of the nineteenth century that research really came to the fore.[26]

It follows that one cannot speak of a Berlin or a Humboldt model in German academic debate in the nineteenth century. On the other hand, foreign observers *did* remark on the crystallization of a particularly German model in the second half of the nineteenth century: the modern research university. During the German Empire, this model became an export success, and a succession of universities in both Europe and North America were reformed along German lines.[27] Johns Hopkins University, founded in Baltimore in the 1870s, became the first American university to explicitly aim for a unity of academic training and scholarly research, embodied for example in its special graduate school. In the last decades of the nineteenth century, a whole series of newly qualified researchers took up posts at other leading American universities and so helped to spread the new model.[28]

Humboldt's Revival in the Twentieth Century

In the very last years of the nineteenth century, the historian Bruno Gebhardt made a highly significant discovery. He found in an archive Humboldt's

25 Paletschek 2001b, 97f.

26 Paletschek 2001b, 96–100.

27 Schalenberg 2002; Edward Shils & John Roberts, 'The Diffusion of European Models Outside Europe', in Walter Rüegg (eds.), *A History of the University in Europe: Universities in the Nineteenth and Early Twentieth Centuries (1800–1945)*, iii (Cambridge: CUP, 2004); Jürgen Osterhammel, *Die Verwandlung der Welt: Eine Geschichte des 19. Jahrhunderts* (Munich: Beck, 2009), 1132–1147.

28 Roy Steven Turner, 'Humboldt in North America? Reflections on the Research University and Its Historians', in Rainer Christoph Schwinges (ed.), *Humboldt international: Der Export des deutschen Universitätsmodells im 19. und 20. Jahrhundert* (Basel: Schwabe, 2001).

unpublished memorandum, 'Über die innere und äussere Organisation der höheren wissenschaftlichen Anstalten in Berlin', and duly published it two years later.[29] Suddenly, Humboldt's short, unfinished text from 1809–1810 was at the centre of the turn-of-the-century debate on German universities. Among those who helped to make the text known, and who launched its author as the forefather and founder of the modern university, particularly notable were the educationist Eduard Spranger and the theologian and science manager Adolf von Harnack. As already noted, the centenary of the University of Berlin in 1910 stimulated interest in Humboldt and prompted much of the veneration that developed seventy-five years after his death, particularly in Prussian academic circles. This Humboldt renaissance should also be seen in connection with general developments in the German Empire in the decades leading up to the First World War. This was a dynamic phase, in which scholarship, education, and historical writings became essential components in the cultural identity that was being formed in the young German nation-state. Humboldt's text could not only serve to promote education and academic freedom; it was of just as much use in justifying the increasing importance accorded to research during this period. The university, one of the institutions that was a source of the greatest pride for the educated German bourgeoisie, became a matter of national importance.[30]

During Germany's dramatic twentieth century, Humboldt—or at any rate, the classical German university model to a greater or lesser extent associated with his name—became an important reference point in discussions about research and higher education. In the 1920s, for example, the minister for culture and education, Carl Heinrich Becker referred to Humboldt, while also arguing that the university must be adapted to the demands of the new age.[31] The Nazis, on the other hand, did not waste much time on academic freedom and personal *Bildung*. The university ideologues of the Third Reich demanded that research and education should be subordinate to the demands of the nation.[32]

29 Paletschek 2001b, 77.

30 Dieter Langewiesche, 'Die "Humboldtsche Universität" als nationaler Mythos: Zum Selbstbild der deutschen Universitäten im Kaiserreich und in der Weimarer Republik', *Historische Zeitschrift*, 290/1 (2010), 53–91; Dieter Langewiesche, 'Humboldt als Leitbild? Die deutsche Universität in den Berliner Rektoratsreden seit dem 19. Jahrhundert', *Jahrbuch für Universitätsgeschichte*, 14 (2011), 15–37.

31 Guido Müller, *Weltpolitische Bildung und akademische Reform: C.H. Beckers Wissenschafts- und Hochschulpolitik 1908–1930* (Cologne: Böhlau, 1991).

32 Notker Hammerstein, 'Humboldt im Dritten Reich', in Rainer Christoph Schwinges (ed.), *Humboldt international: Der Export des deutschen Universitätsmodells im 19. und 20.*

In the wake of the Second World War, a large number of German academics took part in the debate on university ideals, sometimes by explicitly promoting Humboldt or the classical German university as a corrective to the Third Reich. They wished to inject new life into the university, and to protect scholarly integrity, academic freedom, and the pursuit of pure knowledge. An important text from this period was Karl Jaspers's *Die Idee der Universität* (1946).[33] Even in the East German Occupation Zone, the Prussian educational reformer had a role to play. In 1949, the old university in Berlin changed its name to the Humboldt-Universität zu Berlin, named for both Wilhelm and Alexander von Humboldt.[34]

The 1960s saw a radical change in conditions at Germany's universities. Humboldt could still be invoked as a living ideal, as Helmut Schelsky did in his book *Einsamkeit und Freiheit* (1963), which attracted much attention.[35] At the same time, critical and democratic alternatives to the existing university system were launched, and many German traditions, including Humboldt's, were subject to debate.[36] With time, the conception of *Bildung*, which lay at the heart of Humboldt's ideal university, came in for an ever-more pointed philosophical and theoretical critique in many other countries. One influential example was Jean-François Lyotard and his book *La condition postmoderne* (1979).[37]

During the early twenty-first century, the Bologna Process has been the subject of an extensive discussion. The tide of discussion has at times run very high, not least in Germany. International reports created a widespread feeling of crisis, since the German university was universally thought to have been left behind. In the many books published on this theme, Humboldt serves a compensatory purpose, a corrective to the market-oriented mood of the age. Titles such as *Universität ohne Zukunft?*, *Die ungeliebte Universität* and *Humboldts*

Jahrhundert (Basel: Schwabe & Co., 2001); Notker Hammerstein, *Die Deutsche Forschungsgemeinschaft in der Weimarer Republik und im Dritten Reich: Wissenschaftspolitik in Republik und Diktatur 1920–1945* (Munich: Beck 1999).

33 See Johan Östling's chapter in this volume.

34 Reimer Hansen, *Von der Friedrich-Wilhelms-Universität zur Humboldt-Universität zu Berlin: Die Umbenennung der Berliner Universität 1945 bis 1949 und die Gründung der Freien Universität Berlin 1948* (Berlin: Präsident der Humboldt-Univ., 2009).

35 Helmut Schelsky, *Einsamkeit und Freiheit: Idee und Gestalt der deutschen Universität und ihrer Reformen* (Reinbek bei Hamburg: Rowohlt, 1963).

36 Anne Rohstock, *Von der 'Ordinarienuniversität' zur 'Revolutionszentrale'? Hochschulreform und Hochschulrevolte in Bayern und Hessen 1957–1976* (Munich: Oldenbourg, 2010).

37 Jean-François Lyotard, *La condition postmoderne* (Paris: Editions de Minuit, 1979).

falsche Erben send a clear message.[38] This literature has been especially voluminous in Germany, but has its counterparts in many other countries. And it is frequently filled with implicit or explicit references to Humboldt.

At the same time, other fields have developed new approaches that are far removed from the ideals of the classical research university and the Humboldtian model institution. Most influential would be sociology's notion of a so-called *Mode 2*. This much-debated concept was first launched in the mid 1990s to express the new logic according to which knowledge-based processes were said to work.[39] *Mode 1* was a Humboldtian model, in which research proceeded on the basis of subject-specific and researcher-driven processes; knowledge production took place in an essentially closed university, disconnected from the surrounding society. *Mode 2*, on the other hand, was characterized by interdisciplinary approaches, transparency, flexibility, and a new sensitivity to the growing expectations of society, and the demands it placed on the university. Scholarly work had a more problem-solving character and answered the concrete needs of society in a novel way. The concept of *Mode 2* has often been used by Humboldt's critics, who believe that the German university reformer and his conception of the university are lacking in relevance for our time.

The Present Book

This short and selective sketch of the Humboldtian tradition's origin and transformations suffices to show that our initial proposition is correct—that the history of the last two hundred years has been complex. This anthology aims to illustrate and clarify this complex history, not least against the background of the new and partly revolutionary research referred to above. The book also includes contributions that emphasize parts of the legacy which until now have not been considered, despite what at times has been intense research activity in the field.

38 Dorothee Kimmich & Alexander Thumfart (eds.), *Universität ohne Zukunft?* (Frankfurt am Main: Suhrkamp, 2004); Jochen Hörisch, *Die ungeliebte Universität: Rettet die Alma Mater!* (Munich: Hanser, 2006); Christine Burtscheidt, *Humboldts falsche Erben: Eine Bilanz der deutschen Hochschulreform* (Frankfurt am Main: Campus, 2010).

39 See Michael Gibbons, Camille Limoges, Helga Nowotny, Simon Schwartzman, Peter Scott & Martin Trow, *The New Production of Knowledge: The Dynamics of Science and Research in Contemporary Societies* (London: SAGE, 1994); Helga Nowotny, Peter Scott & Michael Gibbons, *Re-Thinking Science: Knowledge and the Public in an Age of Uncertainty* (Cambridge: Polity Press, 2001).

However, the current situation is also taken into consideration. After all, Humboldt is often used as a stick to beat people with in contemporary research and education policy. But what happens if we look past the simplistic, rhetorical posturing of debate and allow the Humboldtian tradition, in all its nuanced richness, to function as a mirror in which to study the questions that engage us today? This book, in other words, will connect the past with the present, and will attempt to define Humboldt's place in the landscape of contemporary research and education. At a time when many universities are undergoing rapid and radical change, questions as to the actual goals and purpose of higher education naturally present themselves. It has always been so in the history of the university, from Wilhelm von Humboldt's own time to the post-war rise of the mass university. In our own time, there is an evermore keenly felt need for self-reflection in the academy. When we turn our gaze to our own activities, the Humboldtian tradition unavoidably comes into clear focus. Or rather, that is the case in many European countries where higher education has been heavily influenced by German ideals formulated at the turn of the nineteenth century.

This anthology was first conceived at a conference that was held in Uppsala in November 2010 to mark the bicentenary of the opening of the university in Berlin, so even this, our book, should be appended to the host of works which followed in the wake of the celebrations. For two days, researchers from various countries presented work that in different ways was concerned with Humboldt then and now. Most of the contributions to this book are based upon these presentations; however, all have been substantially revised and expanded. In addition, we have chosen to add a pair of complementary articles in order to provide a more rounded treatment of the Humboldtian tradition and its transformations.

The book opens with a historical section comprising three chapters that treat aspects of the prehistory and origins of the Humboldtian tradition: what would later be woven together into a single tradition had different origins and different patterns of development. Similarly, many of the central practices and concepts that were gradually introduced into research and education during the eighteenth century were of consequence for the university's general activities and for academics' self-image. The three chapters illustrate in various ways how practices changed and concepts shifted in meaning.

In the first chapter, *Peter Josephson* analyses the importance of publication as an academic merit at German universities during the latter half of the eighteenth century and the early nineteenth century. Today, a university is an institution that conducts both research and teaching. Publications are valued because they are taken as confirmation of the individual researcher's contributions to the advancement of knowledge. However, in the eighteenth century the university was purely a teaching institution. Josephson discusses not only *why* it was desirable that university teachers should write for publication; he is mostly concerned with *how* individuals were induced to take on the role of a writer when they had previously never thought of themselves as anything other than teachers. Josephson believes, as he writes at the conclusion of his chapter, that the discussions inspired by this question at the turn of the nineteenth century constitute previously overlooked background to Humboldt's ideas about the unity of research and education.

Thomas Karlsohn, in turn, concentrates on references to the practice of lecturing in Humboldt's precepts for the University of Berlin. He sketches the background to Humboldt's idea of the extempore lecture, and illustrates the differing practices that characterized the medieval and early modern universities, while demonstrating that Humboldt was not in any sense unique in calling for a new way to conduct lecture-based education. On the contrary, there were important forerunners in the establishment of academic lecturing as a process not bound to follow a predetermined text—as an open-form method of education. Karlsohn also follows the development forwards in time and sketches the history of lecturing up to our own time, with new, technique-driven distance-learning methods, including e-lectures as a central element. The chapter concludes with a brief discussion highlighting the legacy of Humboldt and his age as important starting-points in the present-day debate on higher-education pedagogy.

Claudia Lindén contributes a gendered historical perspective on the late eighteenth and early nineteenth century discussions of a central concept for Humboldt—*Bildung*. It has often been assumed that *Bildung* indicates a form of self-definition that every individual has the potential to achieve. Lindén reads Humboldt alongside Goethe, and shows how, on the contrary, the ideal of *Bildung* found in the works of both writers belongs to a heteronormative and fundamentally misogynist outlook. At the same time as she discusses the differences in Goethe's and Humboldt's views on the relationship between the sexes, and she shows how for both of them *Bildung* is reserved for a markedly male subject, despite the fact that for Humboldt a man could only achieve *Bildung* on the condition that he cultivated the feminine aspects of his character.

The second section of the book brings together four papers that in different ways look at the evolution of the Humboldtian tradition from the nineteenth century onwards. Wilhelm von Humboldt and the university model to varying degrees connected to his name became at various times a reference point in the research and higher-education debate. One point shared by all four authors is their emphasis on the prevailing historical context. It becomes clear that the Humboldtian tradition is not an unchanging, ahistorical process, but is to the highest degree embedded in an intellectual and political context, often in the form of a specific, national educational culture, the authors thus demonstrating how important it is to have an understanding of both the general changes in university tradition and the concrete historical circumstances in which the ideals were fashioned.

Mitchell G. Ash takes as his starting-point how 'Humboldt' has been used in Germany in the recent debates on higher education. He stresses the existence of a kind of Humboldt myth, but argues that we should see this myth as an *invented* tradition that carries a wide variety of cultural meanings. Ash depicts the development of the German mass university, and shows how Humboldt appears in various guises in the debates about university reform after Reunification and about the Bologna Process. In his opinion, an important explanation as to why so many have referred to the Prussian educational reformer during the 1990s and 2000s is that there is a lack of any powerful and convincing alternative. There is as yet no cultural code that can replace Humboldt.

Pieter Dhondt shows in his chapter how Humboldt and the German university have served as a model and a rhetorical resource in Belgium. From the middle of the nineteenth century, the German model exerted a considerable influence on many Belgian scholars. At the same time, the Belgian state was anxious not to allow full academic freedoms. Dhont argues that the form taken by the Belgian university system was thus a compromise between the French vocational education model and the German research university. The First World War affected Belgium badly, and Germany lost its position as the scholarly model; instead, the French model reappeared as an alternative, and, after the Second World War, Anglo-Saxon influences became ever stronger. Nevertheless, by 2000 Humboldt was beginning to appear evermore often in the rhetoric, and Dhondt gives examples from the contemporary Belgian debate on higher education that illustrate how references to Humboldt can sometimes be wholly gratuitous.

In the wake of the Second World War, there was an intensive debate about the universities in the different zones of Occupied Germany. *Johan Östling* analyses the role played by the Humboldtian tradition in these discussions. He especially takes note of the philosopher Karl Jaspers, who considered the ideal of the university in a series of influential articles and speeches. Jaspers and many of his contemporaries were convinced that it was crucial for the survival of German culture that the universities be revitalized. In these debates, Humboldt played a relatively minor role, but Östling asserts that many held fast to what they regarded as the classical German university ideal: academic freedom, the search for pure knowledge, the unity of scholarship, and the conjunction of research and education. Although Humboldt himself was barely mentioned, the tradition that today is often associated with his name was made topical.

One of the greatest changes in higher education in modern times was the growth of the mass university during the 1960s and 1970s. This was a radical process, and one that also invoked protests. In her chapter, *Marja Jalava* discusses how the Left in Finland criticized the dominant, technocratic educational ideal, and in its place promoted Humboldt's *Bildung*-based university as an alternative. She shows how the Finnish Left was part of a larger international movement, but emphasizes also its distinctive national characteristics. For example, the Finnish Marxists not only held up Humboldt and other German idealists as a corrective; the philosopher J.V. Snellman, 'the Humboldt of Finland', was also a source of critical inspiration. By way of conclusion, Jalava discusses the importance of alternative visions of the university.

The book's third and final section presents the work that above all dwells on contemporary issues, investigated in the light of the principles and themes with which Humboldt is usually associated. The history of individual disciplines and their significance for the entirety of higher education; trends in university policy; and the creation of scholarly norms: all are subjects for discussion, as is the modern use of Humboldt as a symbol in the struggle against changing the university into a commercial entity or a politically led authority.

As we have seen, Humboldt often plays an important role in current discussions of the future of our universities. Those who wish to make the university more commercial in scope—frequently drawing inspiration from the OECD—do not hesitate to invoke him. For their opponents he is, of course, a central point of reference. In her chapter, *Susan Wright* discusses this type of rhetoric, and demonstrates how references to Humboldtian principles have become tokens in a complex discursive struggle. She emphasizes that references made by the enthusiasts for market forces to what they insist is Humboldt's legacy conceal a departure from that ideal. Success in argument is dependent on this relationship remaining invisible. Wright considers a number of examples from the Danish public discourse, and she highlights weaknesses in analyses by the opponents of market forces. Most of all, they have failed to grasp how the meaning of the key concept of academic freedom has changed.

Hans Ruin's chapter considers the position of philosophy in the modern university. He opens with a discussion of the discipline's central role in the original, *Bildung* university, a role which was explicated by Kant in *Der Streit der*

Fakultäten and which finds clear expression in the writings of Schiller, Fichte, and Humboldt, among others. Ruin also summarizes the long development of specialization and marginalization that has taken place in the last two hundred years. In a reflection on this development, he refers to his own experience gained in helping to form the discipline of philosophy at Södertörn University in Sweden. Philosophy was regarded as irrelevant by administrators and bureaucrats there. However, once the subject had been established, it was possible to turn their lack of interest to good effect, as it meant there was greater room for manoeuvre. He concludes with a number of reflections in which Humboldt's principles for academic life are related to present-day philosophy and to our contemporary situation: in the current circumstances, Ruin believes we must free ourselves from traditional attitudes towards disagreement within the universities, while at the same time philosophy can offer new meaning.

In her contribution, *Ylva Hasselberg* illustrates the process of norm-building in university life. She demonstrates the use of various depictions of the university, for example those we associate with Wilhelm von Humboldt and Robert Merton, all of which centre on strongly normative assertions about what it means to work with knowledge and to be active at a university—and all of which are now under threat. However, Hasselberg argues for the value of these norms, and not least for the emotionally based ideal of the objectivity of knowledge. Such ideals reproduce and reinforce how we gauge good scholarship, in contrast to the official, political rhetoric of scholarship. Hasselberg takes a historical approach, and demonstrates the various shifts in our understanding of scholarship in modern times. Her chapter concludes with a discussion of economists' understanding of scholarly rationality—which is shown to have a problematic relationship to the classical norms—as well as a critique of the shallow manner in which traditional values are referred to within the academy.

In the last chapter in this section, *Sharon Rider* undertakes an analysis of the pedagogical assumptions and directives that have driven the Bologna Process. It has sometimes been asserted that the demands placed on the universities by this process (for example, that academic education should lead to 'employability') are readily compatible with the ideals of knowledge and *Bildung* that we associate with the Humboldtian model. This opinion is emphatically rejected by Rider. She argues that the Bologna Process is based upon a conception of knowledge and learning that has been rejected by philosophers and pedagogues since antiquity as being not only incorrect, but also harmful. In addition to Humboldt, Rider considers a series of thinkers who have considered the problems of education, and who have all directed sharp criticism at the type of pedagogical ideas that today are being imposed at all levels of higher education throughout Europe.

We would like to thank the foundations whose financial contributions made possible the conference and subsequent work that have given rise to this book: Riksbankens Jubileumsfond, the Swedish Research Council, the Department of History of Science and Ideas at Uppsala University, the Forum for German Studies at Uppsala University, the Forum for Advanced Studies in Arts, Languages and Theology (SALT) in Uppsala, and Letterstedtska föreningen. Finally, we would like to thank Charlotte Merton, who copy-edited the manuscript. Moreover, she translated the introduction and chapter 1, 2, and 6.

Uppsala and Berlin, May 2014
PJ, TK & JÖ

PART 1

Historical Origins

CHAPTER 1

The Publication Mill

The Beginnings of Publication History as an Academic Merit in German Universities, 1750–1810

Peter Josephson

From Dons to Writers

'Is it wholly necessary that professors should become authors, and famous authors at that? This is a new question, and one which awaits answer'.[1] This voice of doubt was Johann David Michaëlis's, writing in the second volume of his *Raisonnement über den protestantischen Universitäten in Deutschland* of 1770. At that time, the profession of university teacher was entering a period of change. The University of Göttingen's leadership had preferred candidates with a good publication record ever since the university was founded in 1737,[2] and it was not long after that the government of Prussia decided to reserve university lectureships for those who had published a minimum of six different works in addition to a doctoral thesis and the further professorial thesis of the type that would later be known as a *Habilitationsschrift*.[3] Governments in other German states had soon adopted similar rules, and adjusted their selection procedures accordingly. Princes and university chancellors in the late eighteenth century consciously strived to recruit the most eminent and renowned writers for their institutions.[4] Johann Adam Bergk drew attention to the change in his *Die Kunst, Bücher zu lesen* from 1799: 'We appreciate the value and the significance of a scholar only from his writings; he who does not write, seems unsuited to the world of

---

1 Johann David Michaëlis, *Raisonnement über den protestantischen Universitäten in Deutschland*, ii (Frankfurt/Leipzig, 1770), 225.

2 Margrit Rollman, *Der Gelehrte als Schriftsteller: Die Publikationen der Göttinger Professoren im 18. Jahrhundert* (Göttingen: Phil. diss., 1988), 65 for example.

3 William Clark, *Academic Charisma and the Origins of the Research University* (Chicago: University of Chicago Press, 2006), 259f.

4 See, for example, Helmuth G. Walther, 'Die Universität um 1800', in Gerhard Müller, Klaus Ries & Paul Ziche (eds.), *Die Universität Jena: Tradition und Innovation um 1800* (Stuttgart: Franz Steiner Verlag, 2001), 27–32 at 31f. Walther's point is that Jena's popularity in the late eighteenth century should be seen as resulting from the university leadership's strategy of recruiting famous writers.

 | DOI 10.1163/9789004271944_003

scholarship today, and when someone applies for a position, one always asks what he has written'.[5]

The introduction of the new system of promotion was one link in a process that in time would fundamentally change the role of the academic. Further measures to encourage professors and lecturers to publish were being considered in many places in Germany by the end of the eighteenth century. Naturally, in the past there had been university teachers who regarded themselves as writers—the academic world seems to have always attracted individuals with ambitions to publish—yet, with the exception of lecture manuscripts and the texts that formed the basis for oral disputations, a professor rarely wrote anything in the exercise of his office. The fact that some university teachers never wrote a single line beyond what the position required had not been seen as a problem. However, during the period considered here, a change took place: from then on, university teachers were not only expected to publish, but to publish extensively.

My aim in this essay is to investigate this change. I would not like to give the impression that historians have overlooked the weight increasingly placed on academic publication: in recent years, William Clark in particular has made a major contribution to the field; and an older study with a closely related focus is Margarit Rollmann's *Der Gelehrte als Schriftsteller* from 1988. However, my ambition here is different from the line taken in previous research. My purpose is not so much to show that there was a desire to induce university teachers to write for publication, or to investigate why, although the reasons for the new regime will be considered: I am more interested in *how* this change was to be achieved and the available *means* with which to implement it. In short, it is not enough merely to state the fact that university appointments became ever more dependent on a record of published writing. This was an important factor in the overall picture—but only one. Other circumstances need to be taken into account as well. How could a professor be encouraged to continue publishing once he had taken up a chair? What incentives to publish new work were there for an academic who found himself in that situation? What means were available to induce someone who had never previously thought of himself as anything other than a teacher to take on the role of author?

It is important here to resist the temptation to be anachronistic. To a modern academic, intent on conducting and publishing research, the German universities of the eighteenth century can appear inhospitable: many libraries were in a miserable state; anything one wrote could be seized or withdrawn by

5 Johann Adam Bergk, *Die Kunst, Bücher zu lesen: Nebst Bemerkungen über Schriften und Schriftsteller* (Jena, 1799), 56.

the censor at any time; and one could be forced to teach from morning to night. I believe, however, that we should be wary of describing these factors as obstacles. The problem with such a characterization is that it reinforces the tendency to take the incentives and driving forces as naturally occurring, when they rather need to be examined in terms of their historical origins.[6] It is as if one were to begin by supposing that academics of all eras have always shared an overwhelming desire to publish, and that the only things preventing them from doing so were wholly external factors. Instead of taking this desire as a given, I wish to bring to light the methods and forms of control by which it was once created and instilled. Towards the end of the chapter, I will ask how the objects of these new publication policies reacted to the new order. In an epilogue, I discuss how the material investigated in the body of the essay helps us to understand the character of the teaching institution Wilhelm von Humboldt intended to create when he set about founding the university in Berlin in 1810.

Money and Honour

The most important reason why we today regard publication as an academic merit is that the university's remit is not only to teach, but also to research. Publications are valued because they document a candidate's contributions to the advancement of knowledge. However, in the eighteenth century the university was purely a teaching institution. A university teacher was considered to have fulfilled his duties when he had taught his way through the received wisdom of his field. Rarely was he expected in addition to this to contribute original ideas or knowledge. Before we proceed, there is therefore reason to ask why it was the case that candidates who wrote and published books were preferred. Why did universities want teachers who had distinguished themselves as authors?

At the start of the eighteenth century, Germany's rulers had little interest in intervening in academic appointments. Recruitment was conducted in the same way as in most guilds. Sons and sons-in-law inherited posts from their fathers and fathers-in-law. In many places, professorial chairs were divided between a small number of families, who together ensured that the line of

6 I will not consider the re-evaluation of writing per se, however; for example, I do not go into the Romantics' fascination with originality and its ramifications for the present context. In the period in question, the written word was to become the most important expression of one's creative ability and originality. This has recently been treated in Thomas Karlsohn *Originalitetens former: Essäer om bildning och universitet* (Gothenburg: Daidalos, 2012).

succession remained unbroken. However, by mid century this model was coming under strain. Across Germany, student numbers had declined in the first half of the eighteenth century—a worry for rulers schooled in the maxims of Cameralist public administration. Universities were not only intended to provide an education for priests and officials, but should also contribute to the general prosperity.[7] Falling student numbers were synonymous with lower revenues, not just for the university teachers, but for the state too. Faced with this situation, the university appointments system came under scrutiny, and, instead of employing relatives, lecturers were recruited who had the qualities needed to attract students.

This is one reason why lecturers who had made a name for themselves as writers were increasingly sought after. It was thought that famous, productive writers would bestow honour and reputation on their university, and thus could also help to improve student numbers. The rector of the University of Göttingen, Ernst Brandes, made the connection in a book published in 1802: 'Much of the renown of a university and its professors in Germany, and almost all the renown that a university and its members enjoy outside Germany, depends upon the published writings of the professors'.[8] When it came to attracting students from neighbouring German states, it was regarded as especially important to have teachers whose books had attracted attention. 'For', as another observer noted, 'it is only through publication that the teacher's reputation is carried far and wide, and attracts new audiences to his lectures'.[9]

The fact that academic promotions were influenced by financial concerns was not something anyone had a mind to conceal. The habit of preferring men who had publications to their name was also justified by the argument that this was good for the local publishing and printing industry; an opinion advanced, for example, in an article in the *Wissenschaftliches Magazin für Aufklärung* in 1786. The anonymous author—who informs the reader that he had served as a government minister and a university dean—regards academic publications as a way for the state to create prosperity:

> Bad times await a university if its teachers only write little, or not at all, and mostly teach from compendia, having to explain others' words and

7 Andre Wakefield, *The Disordered Police State: German Cameralism as Science and Practice* (Chicago: University of Chicago Press, 2009), 49.

8 Ernst Brandes, *Ueber den gegenwärtigen Zustand der Universität Göttingen* (Göttingen, 1802), 189.

9 Carl Villers (Charles de Villers), *Über die Universitäten und öffentlichen Unterrichts-Anstalten im protestantischen Deutschland insbesondere im Königreiche Westphalen* (Lübeck, 1808), 110.

> thoughts.—Who would not rather hear the author himself, rather than his commentator? Furthermore, a vast sum of money comes into the country from the sale of university publications, and therefore it is excellent to see a university whose scholars are industrious writers.[10]

It should be pointed out that it was not just rulers and government officials who adopted an economic perspective on the production of academic papers. The same was true of many in the academic world. Johann David Michaëlis, who himself rejected the idea that publication should become obligatory for university teachers, admitted in a paper from 1770 that it would nevertheless be beneficial from a financial standpoint if he and his colleagues wrote more books.[11] When, in a text from 1802, the philosopher Christoph Meiners defended the still not generally accepted notion that an academic should be active as a writer, it was once more the economic arguments that weighed most heavily. Meiners assured his readers that he certainly believed that a university should be seen as a teaching institution, and that instruction should continue to be given in verbal form. Yet, as he makes clear, that is not enough. 'It is thus not enough for a university to be what it is expected to be. It must also be seen to be so. The flourishing of the schools of higher learning, inasmuch as it is based upon the merits of their teachers, depends most of all upon the literary fame of their teachers'.[12]

Naturally, a long list of publications was not the only merit that gave priority in academic positions. The ability to lecture also counted. University teachers who were known for being good lecturers could, just like their colleagues noted

---

10 Anonymous, 'Bemerkungen über Johann Jacob Mosers Rede, wie Universitäten, besonders in den juristischen Facultät, in Aufnahme zu bringen und darinn zu erhalten—aus den Papieren eines verstorbenen Staatsministers und Universitätscurators' (1786), in Emil F. Rössler (ed.), *Die Gründung der Universität Göttingen: Entwürfe, Berichte und Briefe der Zeitgenossen* (Göttingen: Vandenhoeck & Ruprecht, 1855), 468–483 at 471 (Martin Gierl, 'Die Universität als Aufklärungsfabrik: Über Kant, gelehrte Ware, Professoren als Fabrikgesellen und darüber, wer die universitätshistorisch herausragende programmatische Schrift des 18. Jahrhunderts in Wirklichkeit geschrieben hat', *Historische Anthropologie: Kultur, Gesellschaft, Alltag*, 13/3 (2005), 367–375 identifies the author as Friedrich Phillip Carl Böll, alumnus of the University of Göttingen). Regardless of the identity of the author, his was an opinion that was shared by many in the eighteenth century; see, for example, Johann Lorenz von Mosheim, 'Denkschrift über die Einrichtung einer Academie mit Bemerkungen Just. Henning Böhmers' (1733), in Rössler 1855, 20–27 at 21.

11 Michaëlis 1770, 228f.

12 Christoph Meiners, *Über die Verfassung und Verwaltung deutscher Universitäten*, ii (Göttingen, 1802), 9.

for their publications, contribute to increasing student numbers, and thus university revenues. 'The repute that a good lecture proclaims on every side, that of written fame in far-off lands, flourishes, as does the popularity of the seats of higher learning', wrote Meiners.[13] However, it was occasionally said that rarely was one person well qualified in both of these important respects. The question then was which skill should be more highly valued. As will be seen below, many feared that German rulers preferred famous and popular writers over skilled lecturers.

It is easy to understand why suspicions were aroused. The list of people who first achieved success as authors and subsequently were recruited as university teachers would be a long one. August Gottlieb Meissner—today regarded the father of German crime literature—was appointed to the chair in aesthetics at the German university in Prague after his first successes as a writer in the 1780 s.[14] Johann Gottlieb Fichte—who had no academic qualifications, having been forced to break off his university studies early—was made a professor at Jena in 1794 after several years when he had attracted attention as a writer.[15] Five years before, Friedrich Schiller had been awarded a chair in history at the same university on the urging of Goethe. At the time, Schiller was perhaps Germany's most fêted playwright and poet, on a par with Goethe himself, and undoubtedly a recruitment with special star quality. His inaugural lecture, 'Was heisst und zu welchem Ende studirt man Universalgeschichte', attracted such an audience that they were packed in like sardines to get a glimpse of the famous speaker.[16] Judging by the public interest, his employer, Karl August of Weimar, had done very well by his university when he succeeded in engaging Schiller.

Money Talks

What was the most effective way of inducing those individuals who had become part of the university system to carry on publishing? Why would

13 Meiners 1802, ii. 55.

14 In the inaugural lecture Meissner gave when he took up a chair in Prague, he explained that hitherto he had only communicated with his public in writing, and therefore had little experience of lecturing (see August Gottlieb Meissner, 'Über die Pflichten eines Lehrers und den Unterschied von Schrift und Vortrag' (1786), in id., *Sämmtliche Werke* (Vienna, 1814), xxxvi. 162–171).

15 Anthony J. La Vopa, *Fichte: The Self and the Calling of Philosophy*, 1762–1799 (Cambridge: CUP, 2001), 240ff.

16 Rüdiger Safranski, *Schiller oder die Erfindung des Deutschen Idealismus* (Munich: Carl Hanser Verlag, 2004), 310ff.

someone who had already been made a professor waste time on writing books? A university teacher at the end of the eighteenth century could spend up to six hours a day lecturing, sometimes more.[17] It is easy to understand that there was accordingly little time left over for other tasks. With hindsight, one should perhaps ask if the most effective way to increase the rate of publication would have been simply to free university teachers from some part of this teaching load. When measures to stimulate research are discussed today, a reduction of teaching hours is sometimes advanced as a solution. The success of such initiatives, however, depends on those who are affected by them truly regarding both research and publication as integral parts of their work. The problem was that this attitude did not yet exist at the start of the period under consideration here. Before seats of learning could produce more publications, they were thus forced to produce more writers.

How was this to be done? I have already quoted from an article published by a former government minister and university dean in the *Wissenschaftliches Magazin für Aufklärung* in 1786; in the same article, the author states that rulers should learn to take advantage of the university teachers' basest and most selfish desires, and to use them to more general public benefit. 'Self-interest, rivalry, jealousy—all must be brought to bear if the academy is to rise to the highest levels of renown'.[18] It would not be at all difficult to persuade university teachers to write, were the authorities only to understand how to satisfy their need for flattery and appreciation, and to repay their efforts generously: 'The means needed to set learned quills in motion are not great. Let us only be Maecenas, and we shall not lack for Virgils'.[19]

It was not only rulers and government officials who discussed what should be done to make professors and lecturers produce more books; the dilemma

17 It is not easy to be precise as circumstances varied considerably, and it was partly up to individual lecturers to decide how much they wanted to teach. A full professor lectured a number of hours a day, and drew a state salary for doing so. In Leipzig, for example, the agreed time was four hours (see *'Der Universitäts-Bereiser' Friedrich Gedike und sein Bericht an Friedrich Wilhelm II*, i: *Ergänzungsheft des Archivs für Kulturgeschichte*, ed. Richard Fester (Berlin: A. Duncker, 1905), 87). In addition, a university lecturer could teach *privatissima*, privately tutoring students who paid out of their own pockets: this was very common, being an important source of extra income for lecturers. In his report to Friedrich Wilhelm II, Friedrich Gedike mentioned one lecturer in Helmstädt who gave between ten and twelve colloquia a day (ibid. 8f.). Six hours a day was given as the norm by one author (Anonymous 1855, 469), and in the light of other contemporaneous sources this seems a good educated guess.

18 Anonymous 1855, 472.

19 Ibid. 471; but see ibid. 474.

also engaged representatives of the academic world. Many of those who took part in the debate were of the unspoken opinion that many university teachers would refrain from writing in the absence of a strong external stimulus. Here, economic inducements appeared best suited to the purpose. Much like the anonymous author quoted above, Christoph Meiners thought that university teachers who wrote books in addition to their other duties should receive higher salaries.[20] According to Meiners, this had been done at his own institution.[21] However, the Halle philosopher Christian Garve predicted that higher salaries might have the opposite effect if they made the recipient less dependent on the additional income brought in by critical acclaim of their publications.[22] Another commentator was of the opinion that university teachers were largely motivated by the prospect of continued advancement, and noted that it was therefore important to maintain existing internal hierarchies within the universities. 'The untenured teacher, if this distinction is maintained, will wish to become a full teacher; and will consequently make great efforts, through good lectures, diligence in the classroom, and useful publications, to make a name for himself...This important means of motivating younger aspiring professors will be lost if those distinctions are abolished'.[23] An important part of the solution was thus to maintain, or even strengthen, pay and status differentials between various groups of university teachers, and at the same time to adjust the system of promotions so that it put a greater premium on those qualifications now in demand.

Another strategy used to encourage university teachers to publish was to expose them to greater competition by revoking their monopoly on teaching certain subjects. Friedrich Phillip Carl Böll discussed this tactic in *Das Universitätswesen in Breifen* from 1782, in which he presented his argument in the form of a fictional correspondence between two university teachers. One, V.C., has recently been appointed the dean of a smaller university. His friend, C.B., is now professor emeritus, but previously taught at one of Germany's leading universities. As V.C. is new to his job and keen to attract more students,

20 Meiners 1802, ii. 54f.

21 Christoph Meiners, *Kurze Darstellung der Entwickelung der hohen Schulen des Protestantischen Deutschlandes, besonders der hohen Schule zu Göttingen* (Göttingen, 1808), 20.

22 Christian Garve, 'Ueber Gesellschaft und Einsamkeit' (1800), in id., *Versuche über verschiedene Gegenstände aus der Moral, der Litteratur und dem gesellschaftlichen Leben* (Breslau, 1800), iv. 228.

23 Anonymous, 'Über das Prorektorat auf Universitäten und einige andere akademische Gegenstände', *Berlinische Monatsschrift*, 1795, 220–242 at 231.

he asks C.B. for advice. The latter duly presents a plan. According to C.B., there is much to be gained by inducing university teachers to compete with one another for their audiences' favour. He therefore encourages V.C. to repeal the regulation still in force at his university that prohibits professors from stealing a subject already taught by another professor. In fact, it is only to the good if several teachers lecturer on the same subject, since they must all exert themselves to the utmost so as not to lose their students. In addition, tougher competition would force them to test new ways of increasing their popularity:

> The teachers will be up and doing, they will drive one another on. Soon they will start to write, when previously they thought producing books was not their business. In my mind's eye I see the compendia, methods, observations, amoenitates, commentaries; by which grows another branch of commerce, the all-important book trade, and that which before only vegetated is now revived; *fiat lux* [let there be light!]. Do not listen, your Excellency, to the lamentations of a few academics; for it is a sure consequence that some will lose all their audience and become almost consumed by envy and jealousy; the whole, however, gains immeasurably.[24]

A noted writer was usually a sought-after teacher; students would flock to him. Those who did not publish would not survive in the new system. The old style of teacher, many of whom passed on their knowledge only in the form of lectures, appears here to have been an endangered species.

Böll's account was probably intended as a satire; however, good satire is never far from the reality it refers to—it works by recognition, after all. And the fact remains that Böll uses C.B. to recommend a model that had already been adopted in several places. At the close of the eighteenth century, the university with the largest number of famous writers among its staff was Göttingen. There the professors' preferential claim to their subjects had been abolished at an early stage, and exactly the sort of competition C.B. called for was encouraged.[25] However, one thing that did not appear in Böll's description was that the system had proved to have unwelcome side-effects. University teachers

24 Friedrich Phillip Carl Böll, *Das Universitätswesen in Briefen* (n.p., 1782), 62. A slightly longer account is given in Martin Gierl 2004, 367–375, who also cites some of the same quotations used here, although to different ends as his purpose is not to analyse publication as an academic merit.

25 It was clear to anyone familiar with the academic world at the time that it was Göttingen that Böll had chosen to satirize.

who lectured on the same subjects, and therefore competed for the same students, were quick to quarrel. August Ludwig Schlözer and Johann Christoph Gatterer, who both taught universal history, both openly accused the other of attempting to promote himself and his courses by anonymously reviewing his own textbooks in various quarters,[26] an example that is interesting not only because it shows one way to proceed should one wish to manipulate the system, but also because, whether the accusations were true or false, they say something about the importance university teachers attached to their books as public relations material.[27] A book was a shop-window where professors and lecturers paraded themselves before the young men about to embark on a course of study. It was important to make a good impression, otherwise there was always a risk that the audience would disappear to listen to someone else—and with them went the money.

The Measure of the Man

Successful control by economic means supposes that one understands what it is that is to be controlled. That ought to be more obvious today than perhaps at any earlier time. In recent decades, a wide range of supposedly sophisticated tools have been employed to assess the extent to which work carried out at universities complies with the criteria for efficiency and quality set down by grant-giving authorities. These days, it is not only noted how many articles and books a researcher has written: the journals and publishing houses who have published the work are ranked according to their impact, while citation indices show how often and in which contexts other researchers have referred to the work. Moreover, departments and faculties spend ever more time on quality evaluations, on the directive of governments and funding agencies. These evaluations not only provide information about how organizations work; they also contribute to the shaping of those same organizations, given that the results largely determine how educational resources and research

26 It is possible to follow the course of the conflict through the two authors' salvos: August Ludwig Schlözer, 'Species facti', in id., *Vorstellung seiner Universal = Historie* (Göttingen/Gotha, 1772), 402–416; and Johann Christoph Gatterer, *Antwort auf die Schlözersche Species Facti* (Göttingen, 1773), especially 41.

27 The custom of reviewing one's own work seems to have been quite common at this point, and there are plenty of examples of writers who accused one another of having done so. The issue is discussed in detail in Frank Donoghue, *The Fame Machine: Book-reviewing and Eigteenth-century Literary Careers* (Stanford: SUP, 1996).

funds are allocated.[28] When describing this development, Cris Shore and Susan Wright turn to Foucault and his analysis of Bentham's panopticon: 'The rationality of audit thus appears similar to that of the panopticon: it orders the whole system while ranking everyone within it. Every individual is made acutely aware that their conduct and performance is under constant scrutiny'.[29]

From here, it can seem that the step back to the late eighteenth century is a very long one. To what extent were academics then subject to inspection and supervision? Two hundred years ago, who counted how many books a professor had produced?

The truth is that even before the period in question, broadly systematic evaluations of the activities of Germany's universities were already being carried out, for until the close of the eighteenth century German governments conducted visitations, inspecting the organizations and institutions which were under the jurisdiction of the state and the Church, schools and universities among them.[30] An academic visitation was usually carried out by a visitor appointed by the ruler; on arriving at the university, he would demand that all professors should present themselves for examination on how they fulfilled the duties of their position. The procedure had been introduced in the sixteenth century, initially to ensure that the education on offer did not deviate from correct Christian doctrine. However, over time the visitors' brief changed. In *Academic charisma and the origins of the research university*, William Clark examines visitation books of various periods, and discerns what he describes as a transition from 'legal–ecclesiastic' to 'political–economic' terms of reference.[31] Clark's final example comes from a visitation of the University of

28 See Marilyn Strathern (ed.), *Audit Cultures: Anthropological Studies in Accountability, Ethics and the Academy* (London: Routledge, 2000); and Michael Power, *The Audit Society: Rituals of Verification* (Oxford: OUP, 1997), 98–104.

29 Cris Shore & Susan Wright, 'Coercive Accountability: The Rise of Audit Culture in Higher Education', in Strathern 2000, 76f.

30 Ernst Walter Zeeden & Peter Thaddäus Lang (eds.), *Kirche und Visitation: Beiträge zur Erforschung des frühneuzeitlichen Visitationswesen in Europa* (Stuttgart: Klett-Cotta, 1984); Gerald Strauss, *Luther's House of Learning: Indoctrination of the Young in the German Reformation* (Baltimore: Johns Hopkins UP, 1978), 249–267.

31 Clark 2006, 340–372; see also William Clark, 'On Ministerial Registers of Academic Visitations', in Peter Becker & William Clark (eds.), *Little Tools of Knowledge: Historical Essays on Academic and Bureaucratic Practices* (Michigan: The University of Michigan Press, 2001). On the use of visitations for secular ends, see Zeeden & Lang 1984, 17; and Martin Honecker, 'Visitiation', *Zeitschrift für evangelisches Kirchenrecht*, 17 (1972), 337–358 at 341f.

Wittenberg in 1789. The visitation book consists of four columns: the first documented the respondent's qualifications as a published writer (how much he had published in the previous year, and on what subjects); the second column recorded the size of the audience at his lectures and his reputation as a teacher; then followed details of his income and private finances (how much he earned from teaching, and whether he had any additional income); and the fourth and last column seems to have been reserved for information which did not fall into the previous three categories, but which might nevertheless be of interest to the ruler—here the visitor could set down hearsay about the man's morals and character, as well as information about any doings not directly connected to his position. According to Clark, it is reasonable to suppose that records of this type were used when deciding the salary a university teacher should receive from the state.[32]

It is easy to see in the visitation an early forerunner of today's panoptical regulation of the universities. I could not agree more when Clark emphasizes that records of the type he analyses do not just passively register an academic reality that was separate from the evaluation method itself; instead, visitations reminded university teachers that their activities were subject to control, and the kinds of achievements that were expected and rewarded by their employers. In this sense, they created and propagated a certain level of discipline.[33]

However, there is a paradox here. Clark and I both regard the second half of the eighteenth century as a period when governments across the German-speaking region attempted to rationalize the activities of their universities. Given this, one would expect the visitations to intensify towards the end of the century. Yet this did not happen. The fact is that inspections of this sort took place more and more infrequently from mid century onwards, and they soon ceased altogether. The University of Wittenberg visitation in 1789 was to be the last ever conducted by a German government.

Where did rulers and patrons then gather their information about what was going on at their universities? One possible reason why visitations were no longer thought indispensable may have been that by the end of the eighteenth century there were other sources which collected and collated the very information the authorities were looking for. One illustrative example are the books printed around that time with the express intention of guiding young men in their choice of university: works with titles such as *Göttingen. Nach seiner*

32 Clark 2006 makes vague mention of the way in which the information was used, and the extent to which the university teachers who were assessed were made aware of the criteria used to judge their efforts.

33 Ibid. 340 & 372 for example.

eigentlichen Beschaffenheit zum Nutzen derer, die daselbst studiren wollen ('Göttingen: on its true nature, for the benefit of those intending to study there') or *Vertraute Briefe über Halle vorzüglich die Friedrichs-Universität daselbst. Für Eltern und Jünglinge, welche die Academie daselbst beziehen wollen* ('Confidential letters about Halle, excellently describing the Friedrich University there: For parents and young who wish to attend the academy').[34] Books of this type always included the same sort of information. They gave a general picture of the place of study, and supplied the reader with information about the university and any associated institutions. More importantly for us, they also often listed longish evaluations of individual university teachers. Thus in the Halle and Göttingen guides, all the professors and lecturers there were rated for things such as teaching skill and the price of attending their colloquia, as well as their personal character and moral standing. Parents who wanted their sons to attend university could thus obtain all the information that might conceivably be of interest. Who were the best and most popular lecturers? Who allowed poorer students to attend their colloquia free of charge? Who had distinguished themselves as writers?

Another source of information that should be mentioned was the system of published reviews. In the same period that saw the publish-or-perish model introduced to the universities, there was a dramatic increase in the number of review journals and other academic periodicals.[35] Despite the parallel increase

34 Carl F.A. Hochheimer, *Göttingen: Nach seiner eigentlichen Beschaffenheit zum Nutzen derer, die daselbst studiren wollen* (Lausanne, 1791); Bogatsch, *Vertraute Briefe über Halle vorzüglich die Friedrichs-Universität daselbst: Für Eltern und Jünglinge, welche die Academie daselbst beziehen wollen* (Giebichenstein, 1798). There has yet to be a systematic study of the genre, despite guidebooks of this type having been common, especially for the larger universities. See, for example, anonymous, *Zeichnung der Universität Jena: Für Jünglinge welche diese Akademie besuchen wollen* (Leipzig, 1798); Anonymous, *Interessante Bemerkungen über Göttingen als Stadt und Universität betrachtet: Für Jünglinge, die dort studieren wollen aber auch für andere zur Belesung* (Glückstadt, 1801); Johann Georg Friedrich Papst, *Gegenwärtiger Zustand der Friedrich Alexanders Universität zu Erlangen* (Erlangen, 1791); and H.P.F., *Freymüthige Briefe an Herrn Grafen von V. über den gegenwärtigen Zustand der Gelehrsamkeit der Universität under der Schulen zu Wien* (Frankfurt/Leipzig, 1775). For a contemporaneous discussion of how this literature may have influenced lecturers' teaching methods, see Friedrich Christian Laukhard, *Briefe über Jena* (Frankfurt/Leipzig, 1793), ixff.

35 Hans Erich Bödeker, 'Journals and Public Opinion: The Politicization of the German Enlightenment in the Second Half of the Eighteenth Century', in Eckhart Hellmuth (ed.), *The Transformation of Political Culture: England and Germany in the Late Eighteenth Century* (Oxford: OUP, 1990), 423–445 at 427–435.

in the number of published books, the number of titles was still not so great as to prevent the *Göttingische gelehrte Anzeigen* long maintaining its ambition to review everything that might be thought of interest. Certainly, observers regarded this kind of publication as being of great importance. It was thought that an author who was praised in the most respected journals would find it easier to advance through the university ranks. According to Johann David Michaëlis, it was often such journals that the rulers used when deciding who should be awarded a chair. He admitted that this could be a reasonable way of proceeding, as long as one found out who had written the review in each case and questioned them further about the candidate. 'A book that costs a couple of thaler or ducats you can buy on the strength of his word; but when it is the choice of a professor, one is well-advised to learn in advance who the reviewer is on whose word one wants to choose. This is generally not difficult, and then, if he is a trustworthy man, he can give counsel when he has ceased to be anonymous'.[36] A review journal could in this way serve as a point of contact; one which continually evaluated and pronounced judgement on authors employed by the universities, and which, when needed, could supply rulers with supplementary information. However, Michaëlis feared that the journals sometimes had altogether too great an influence on appointments. According to him, employers did not always consider carefully enough the source of these opinions, as long as they had appeared in a well-respected journal and reviewers in several journals concurred.[37]

A review provided information about the tenor of a work, and sometimes its quality too; other sources were available for those who were more interested in quantitative measures. In Friedrich Christian Laukhard's novel *Annalen der Universität zu Schilda*, when the Duke of Schilda and his secretary, Schneller, discuss whom to appoint as a new professor, they turn to Johan Georg Meusel's dictionary of authors, *Das gelehrte Deutschland*. First printed in 1767, Meusel's tome had run to five revised editions by 1834, and contained information on every academic author in Germany during the period in question.[38] The articles were arranged in alphabetical order by surname, and gave

36 Michaëlis 1770, 433f.

37 Ibid. 432.

38 Friedrich Christian Laukhard, *Annalen der Universität zu Schilda, oder Bocksstreiche und Harlekinaden der gelehrten Handwerksinnungen in Teutschland; zur Auflösung der Frage: Woher das viele Elend durch so manche Herren Theologen, Aerzte, Juristen, Kameralisten und Minister?* ii (n.p., 1799). *Das gelehrte Deutschland* was published by the Göttingen professor, Georg Christoph Hamberger, until 1773, at which point Johann Georg Meusel became editor.

basic information about the author's academic title and position, followed by a chronological list of all his publications. Anyone who wished to could thus compare two or more authors, scrutinizing their choice of subject, or any differences in publishing frequency—exactly like the characters in Laukhard's novel. The fact that Meusel's dictionary was regularly revised and republished also meant that one could follow how an author's writings developed over time.

Taken together, these classes of material formed a virtual archive that made it possible to examine the qualifications of almost every German-speaking university teacher before 1800 (a possibility that remains open to us today, of course). A family thinking of sending their sons to university could compare various institutions beforehand, along with the reputations and supposed proficiency of their prospective teachers. A ruler who was about to fill a vacant chair could use reviews and relevant reference works to rank candidates according to the qualities that interested him. The fact that many involved in the debate claimed that this was how many rulers set about university appointments meant that university teachers had good cause to be especially careful about how they were judged in such contexts. Who was said to have the most students? Who was the most acclaimed writer? How extensive was one's own publication list when compared to one's colleagues?

Anyone who knows he is being examined and observed, examines and observes himself. Like the panoptical instruments used today in the university sector, the eighteenth century's admittedly more primitive tools doubtless offered many opportunities not only for employers to assess their employees, but also for employees to assess themselves and to compare their own efforts with others'. This, combined with knowledge about which competences would be rewarded by the market, was probably an important precondition for the changes seen in the role of university teachers at the time. The notion that a university teacher should also actively publish was incorporated into the professional self-image that dons constructed for themselves. In *Appell an meine Nation* from 1795, Johann Georg Heinzmann said that the increase in the number of academic publications around that time was due to the sharpening competition between university teachers. 'To place themselves in the first rank of faculty scholars, to attain a position; this constitutes the main endeavour of the majority of academic authors'.[39]

39 Johann Georg Heinzmann, *Appell an meine Nation über Aufklärung und Aufklärer, über Gelehrsamkeit und Schriftsteller, über Büchermanufakturisten, Rezensenten, Buchhändler, über moderne Philosophen und Menschenerzieher; auch über mancherley anderes, was Menschenfreiheit und Menschenrechte betrifft* (Bern, 1795), 122.

Survival of the Fittest

The introduction of bibliometric evaluations would soon bring a fundamental change to the university teachers' working conditions. Even before then, at the end of the eighteenth century, it appeared that professors who published extensively were in pole position compared to those who still preferred to share their knowledge in the form of lectures. Carl Friedrich Elsässer, professor of law and sometime pro-rector of the University of Erlangen, discussed how the academic profession had changed in his *Einige Bemerkungen über akademische Gegenstände* of 1793. Elsässer described a situation in which academic posts and privileges were increasingly reserved for university teachers who preferred writing over their teaching obligations. Since the salary a lecturer received from the state did not cover more than the most basic of living costs, there were few who could afford to resist this development. The university teacher of the old sort, whose hallmark was the well-delivered oral lecture, inevitably appears as a pitiable figure.

> Our man does not have time with which to court the applause of the general literary public with his writing; he receives only scanty allowances, or none at all, because the higher authorities show regard for their more productive honoraria; and when at last his health is spent, so too are his perquisites. He now sits, weakened, and still far from old man, with a salary of about five hundred Thaler, while his book-wright colleague, who contributes nothing, or woefully little, to the good of the individual academy he serves, merely by his writings (perhaps calculated not by their quality, but by their weight) receives like prebendaries a sinecure of twelve hundred Thaler.[40]

Publish or perish. Elsässer would probably have interpreted this maxim literally: the more time a university teacher spent developing his teaching skills, the lower his income would fall, and his estimation in the world with it. If he did not write, he would languish away slowly but surely.

---

40 Carl Friedrich Elsässer, *Einige Bemerkungen über akademische Gegenstände* (Stuttgart, 1793), 16. Elsässer's opinion may well have been coloured by the fact that he was at the University of Erlangen, where lecturer's salaries were relatively low (see Notker Hammerstein, 'Die Universitätsgründungen im Zeichen der Aufklärung', in Peter Baumgart & Notker Hammerstein (eds.), *Beiträge zu Problemen deutscher Universitätsgründungen der frühen Neuzeit* (Nendeln/Lichtenstein: Jacobi Verlag, 1978), 263–298 at 286f and 289).

In the light of this development, how did contemporaries envisage the academic future? Which character traits was it thought the system encouraged and rewarded? Max Weber formulated a principle that is worth recalling here: 'Every order of social relations, without exception and however constituted is, if one wishes to evaluate it, ultimately to be examined with respect to the human type for which it, by way of external or internal (motivational) selection, optimises the chances of becoming the dominant type'.[41] Two hundred years ago, what sort of person was thought to be favoured by the *publish or perish* system? In what ways did the new type of university teacher distinguish himself from the old, other than by writing more books?

One of the objections raised against the new system was that it was rarely the best and most qualified writers who were rewarded with the sort of fame that gave preference in academic appointments. A university teacher who wanted to communicate with a larger audience than a lecture hall would hold, and on those grounds decided to write a book, was forced sooner or later to make contact with a publisher and persuade him to accept the text. Herein lay the problem. The publisher's primary interest was not the advancement of knowledge, but to sell books. Michaëlis touched on this issue too in his *Raisonnement über den protestantischen Universitäten in Deutschland*. He felt that from a commercial viewpoint it was, if anything, a disadvantage if a book was notable for its learning and thoroughness, 'since the reasoning capacity of most book-buyers is so superficial that it is very possible that a book, because it is too learned or can only be of use to a few, will not find enough buyers to pay for the cost of printing'.[42] In the long term, the habit of appointing men who were successful writers to professorial chairs would lead to competent, learned scholars being shut out of the universities. Even the most distinguished would be glad merely to see their work in print at all. 'Humble scholars these days seldom find a publisher for their work', wrote Johann Georg Heinzmann.[43]

A more common objection to these bibliometric evaluation criteria *avant la lettre* was that they seldom favoured talented and enthusiastic university teachers; on the contrary, most critics agreed that people who had a reputation

41 Max Weber, 'Der Sinn der "Wertfreiheit" der soziologischen und ökonomischen Wissenschaften' (1917), in id., *Gesammelte Ausgabe zur Wissenschaftslehre*, ed. Johannes Winckelmann (1968; Tübingen: Mohr Siebeck, 1988), 517.

42 Michaëlis 1770, ii. 236.

43 Heinzman 1795, 128; see also, for example, anonym, 'Ueber die Ursachen einiger Mängel, die sich in der deutschen Litteratur hervorthun', *Neue Deutsches Magazin*, 1 (1801), 238–266 at 238–243; and Friedrich Christoph Jonathan Fischer, *Geschichte des teutschen Handels* (Hanover, 1792), iv. 472.

as important writers were, as a rule, less successful, or simply dreadful, as lecturers. 'Thus with a choice grounded in mere literary reputation some universities brought down on themselves a shining burden', wrote the mathematician Ludewig Thilo in a piece from 1809.[44] The same view was put by A.G. Walch in the *Allgemeine deutsche Bibliothek* in 1782. Walch declared that he shared the opinion that many great writers were awful teachers, often because 'a man who is in the pay of a bookseller, or who seeks a reputation as an author, gladly makes this incidental trade his main occupation, and makes light of the duties of his position, and thus cannot devote all his industry and attention to the education of his charges'.[45] For one or another reason, many were united in believing that the presence of authors in the universities' ranks came at the expense of teaching.

Few, if any, could speak on this question with the same authority as Friedrich Gedike. Quite apart from the fact that Gedike had long been in charge of *Berlinische Monatsschrift*, the Enlightenment periodical bar none, in the 1780s he had been the director of the Oberschulkollegium in Berlin, and in that capacity was ultimately responsible for the recruitment of staff to the Prussian universities. Having left this post in 1789, he toured no fewer than fourteen universities outside Prussia. He noted down his observations in a book that he duly edited on his return and delivered to the Prussian king. On his travels, Gedike had arrived at the University of Jena the day before Friedrich Schiller's installation as professor of history. Gedike described how he was among the audience when 'the famed theatrical poet' delivered what would become his classic inaugural address on universal history. Yet the traveller was not a whit impressed by what he saw and heard. Schiller, he wrote in his report for the king, 'read everything verbatim, in a pathetic declamatory tone that was evidently unsuited to the simple historical facts and geographical notes he had to deliver.... The allure of novelty, the desire to see a famous theatrical poet now at the lectern in a wholly new situation, would likely have led most of such a large body of listeners to gather'.[46] Gedike, it is true, was hardly likely to be surprised

---

44 Ludewig Thilo, *Grundsätze des akademischen Vortrags: Ein Beytrag zur Aufdeckung herrschender Universitäts-Mängel* (Frankfurt/O, 1809), 31.

45 A.G. Walch, 'Vorschläge zur Ziehung und Bildung brauchbarer Lehrer in öffentlichen Schulen, welche von einem Schulmanne in der Kloster- und Stadtschule zu Holzmünden an der Weser erst versucht und ausgeführt, nun aber durch den Druck bekannt gemacht worden sind', review, *Allgemeine deutsche Bibliothek*, 48/2 (1782), 563–569 at 566.

46 Gedike's report was published in its entirety much later (see Richard Fester, *'Der Universitätsbereiser' Friedrich Gedike und sein Bericht an Friedrich Wilhelm II* (Berlin: A. Duncker, 1905), 84).

by the quality of Schiller's lecture. Even before his visit to Jena, he had already criticized the practice of preferring famous writers for university teaching posts. A couple of years earlier he had tackled the subject in a paper on excellence in lecturing: 'The most famous writers are too often the worst lecturers, and choosing a public teacher solely by his reputation as an author is as unfortunate as choosing a bride by her portrait. In both cases, the chooser all too often finds himself deceived'.[47]

Consequently, it was a common belief that universities across Germany would eventually be drained of scholarly and teaching competence if they continued to give priority to famous and popular writers. The new kind of university teacher was driven not so much by a care for the students at his lectures, as by a desire to address anonymous book-buyers. The system was regarded, if anything, as contributing to the moral corruption of the academic world. University teachers who took the new principles to heart were easily taken hostage by the market: people ruled by a lust for honour and profit would find it tempting to tailor the presentation of their scholarship to the tastes of the uneducated masses. One question which has not been touched on here, but which would be interesting to investigate further, is the extent to which the critics succeeded in mobilizing resistance to these anticipated developments. For example, it would be interesting to know the extent to which learned journals and reviews were regarded as tools to frustrate the supposed adaptation of scholarship to market forces in the eighteenth and nineteenth centuries. It is not unreasonable to think that such an investigation could help to illuminate the origins of modern expert review procedures and the peer review process—qualitative evaluation systems that require scholarly perspicacity, and which today are being dismantled and replaced with bibliometry.

Epilogue

In May 1809, Wilhelm von Humboldt took up the post of director of the department of education in the Prussian ministry of culture, where in the months that followed he would lead the work to found a university in Berlin. A short time later, Humboldt wrote a memorandum in which he summarized the principles that he believed should guide and form any institute of higher education

47 Friedrich Gedike, 'Einige Gedanken über den mündlichen Vortrag des Schulmanns' (1786), in id., *Gesammelte Schulschriften* (Berlin, 1789), i. 381–419 at 383 discusses teaching at the Gymnasium in the first instance, but adds that the same approach can be taken in university teaching too.

worthy of the name. The text, 'Über die innere und äussere Organisation der höhern wissenschaftlichen Anstalten in Berlin', today has the status of a classic. Anyone who wishes to understand the ideas that accompanied the growth of the modern research university returns again and again to this document. Among its more acclaimed ideas is Humboldt's notion of the nature of the relationship between research and teaching. Although Humboldt himself never wrote in terms of *Einheit von Forschung und Lehre* (the unity of research and teaching, an expression that is often attributed to him), it still summarizes his general position.

Given the concerns outlined above, this unified vision appears in something of a new light. It has been suggested that in the period before Humboldt's appointment, there had been fears that universities would increasingly be populated by teachers who would shirk their duties to their students in order to retreat into writing. While some worked themselves to the bone in the lecture-rooms, others were wholly taken up with their publications. Humboldt based his argument on a similarly defined problem. Towards the end of the memorandum, he addresses the problem of university teachers who conduct research but fail to understand the value of teaching. Scholarship, he continued, had in Germany always been as favoured at its universities *and* its academies, and this was not by chance:

> Science and scholarship have been advanced as much—and in Germany, even more—by university teachers as by members of academies. University teachers have made these contributions to the progress of their disciplines by virtue of their teaching appointments. For unconstrained oral communication to an audience, which includes a significant number of intelligences thinking in unison with the lecturer, inspires those who have become used to this mode of study just as surely as does the peaceful solitude of a writer or the less institutionalised activities of the members of an academy.... University teaching is moreover not such a strenuous affair that it should be regarded as a distraction from the calm needed for research and study; it is, rather, a help to it. At every large university, there are always men who devote themselves entirely to solitary study and research.[48]

A professor who locks himself in his rooms to research, and prefers to share his knowledge in writing rather than in lectures, in Humboldt's perspective was

48 Wilhelm von Humboldt, 'On the Spirit and Organisational Framework of Intellectual Institutions in Berlin', *Minerva* vol. 8 (1970), 243–250 at 247f.

not only a poor teacher, but also a poor researcher. Humboldt's intention seems to have been to counteract what many of his contemporaries saw as a growing division of labour between two types of academic: those who taught, and those who researched and wrote books. One could also say that he wished to integrate in one and the same profession two academic activities that in preceding decades had increasingly been placed in opposition to each other.

CHAPTER 2

On Humboldtian and Contemporary Notions of the Academic Lecture

Thomas Karlsohn

Introduction

Wilhelm von Humboldt's memorandum from 1809/1810 on a new Berlin university is frequently seen as a manifesto for the modern university.[1] In this short text a number of principles can be inferred that were subsequently taken as fundamental aspects of teaching and research. The freedom of teachers to choose a research topic without interference is one such principle. Another is the unrestricted scope for students to seek out the professor whom they consider suitable. Yet another is the right to teach, unfettered by constraints or external control.

These are familiar fundamentals that have remained topical to this day in the debate on research and educational policy. However, the same degree of attention has not been devoted to the fact that Humboldt also touched on actual university teaching practices, including the function of the academic lecture. He mentioned the subject in a brief passage, where he advanced the notion of oral presentation as an invigorating, dialogic process. At one point he wrote of 'unconstrained oral communication to an audience, which includes a significant number of intelligences thinking in unison with the lecturer'.[2] Humboldt believed that such teaching could involve the students in the knowledge process, and transform them into autonomous educational subjects. During the lecture they would not receive the subject matter passively, but think in unison with the teacher.

Why did Humboldt mention lecturing, and how should the tenor of his remarks be understood? In the following, I interpret his views in light of the situation in which they were formulated. There are a couple studies in which in some ways comparable interpretations can be found, but my contribution includes a more detailed discussion and more source material than

1 Wilhelm von Humboldt, 'On the Spirit and Organisational Framework of Intellectual Institutions in Berlin', *Minerva* vol. 8 (1970).

2 Humboldt 1970, 247.

 | DOI 10.1163/9789004271944_004

previously considered.[3] At the same time, I follow a longer line of historical development that has not been much considered before. It is only against the background of the past that Humboldt's reflections become fully understandable.

My primary, but not exclusive, focus is on the German countries, as it was there that several crucial ideas and practices arose. The elements of my account are scattered throughout the primary sources and earlier research that I highlight and comment on. In addition to this contextualization, I also briefly introduce the possibility of interpreting Humboldt's remarks—and of university lecturing in general—against the background of media history, an interpretation not previously attempted. One important interpretative key to a historical understanding of academic teaching practices lies in the different information technologies predominant in each given period of time. In the case of Humboldt, the key was the new dynamics between oral and written communication that evolved when the printed word became an increasingly dominant aspect of our culture.[4] Here I will limit myself to some short indications of possible areas of further research.

The chapter culminates in a reflection on the current pedagogical challenges facing higher education, mainly those brought by digital technology. In this section, Humboldt and his contemporaries are actualized and put in relation to today's developments in a way not previously attempted. Their formulations with respect to the academic lecture do in fact pose a question directed at our own times. What meaning do these inherited practices and notions hold for us, when digital technology is changing the nature of university teaching?

3 See Hans Jürgen Apel, 'Freier Vortrag vor mitdenkenden Köpfen': Geschichte und Zunkunft der Vorlesung, *Forschung & Lehre*, 2 (1998), 60ff; Hans Jürgen Apel, *Die Vorlesung: Einführung in eine akademische Lehrform* (Cologne: Böhlau, 1999), especially 19–38; and Hans Wenke, 'Die Vorlesung in Vergangenheit und Gegenwart des akademischen Unterrichts', in id., *Hochschulunterricht im Wandel: 4 Vorträge, gehalten auf der konstituierenden Sitzung des 'Arbeitskreises für Hochschuldidaktik am 7.7.67 in Heidelberg* (Göttingen: Otto Schwartz, 1967), 11–36. Apel and Wenke do not, for instance, discuss Schelling's place in the development of the academic lecture.

4 For a discussion of ideas about oral and written communication in the eighteenth century, see Paula McDowell, 'Mediating Media: Towards a Genealogy of "Print Culture" and "Oral tradition"', in Clifford Siskin & William Warner (eds.), *This Is Enlightenment* (Chicago: University of Chicago Press, 2010).

The Middle Ages and Later

Since its beginnings, the university has embraced lecturing as a central practice.[5] It originated in the sermon, preached from *ex cathedra*, the bishop's throne on its dais, and in academia we still use the term chair for a professorship.[6] In the Middle Ages, *lectio* was practiced—the reading aloud of a given text and its exposition in accordance with the tradition that was to be conveyed. In common with the priesthood, the principle task of the professors was to transmit in Latin a body of canonical knowledge. The statutes of the universities specified in detail the content of the curricula that the teachers had to impart.

A key step in oral transmission is systematization and illustration, activities that then took a variety of forms: there was scope for a university teacher to read selected parts of a text aloud, including the comments it contained; he could also interleave reading and quotations with exposition, and, on occasion, various types of aids such as images were employed.[7] The students' task was to assimilate the subject matter, either by dictation or by following the professor's reading in a text in front of them. This was followed by repetitions and disputations. In some instances, a teacher–student dialogue was held during the actual lecture.

---

5 The literature specifically treating the early history of the academic lecture is rather sparse, but there are some valuable works where the subject is treated: William Clark, *Academic Charisma and the Origins of the Research University* (Chicago: University of Chicago Press, 2006); Anthony Grafton, 'Teacher, Text, and Pupil in the Renaissance Class-Room: A Case Study from a Parisian College', *History of Universities*, 1 (1981); Kurt Johannesson, 'Den akademiska föreläsningen', in Arne Jönsson & Anders Piltz (eds.), *Språkets speglingar. Festskrift till Birger Bergh* (Lund: Skåneförlaget, 2000); Bo Lindberg, 'The Academic Lecture: A Genre in Between', *LIR. Journal*, 1 (2011); Arno Seifert, 'Das höhere Schulwesen: Universitäten und Gymnasien', in Notker Hammerstein (ed.), *Handbuch der deutschen Bildungsgeschichte*, i: *15. Bis 17. Jahrhundert: Von der Renaissance und der Reformation bis zum Ende der Glaubenskämpfe* (Munich: Beck, 1996); Rudolf Stichweh, *Der frühmoderne Staat und die europäische Universität: Zur Interaktion von Politik und Erziehungssystem im Prozeß ihrer Ausdifferenzierung (16.–18. Jahrhundert)* (Frankfurt am Main: Suhrkamp, 1991), 285; Rudolf Stichweh, *Wissenschaft, Universität, Professionen: Soziologische Analysen* (Frankfurt am Main: Suhrkamp, 1994), 232; and Wenke 1967, 11–36.

6 Clark 2006, 69.

7 See Edward Grant, 'Science and the Medieval University', in James M. Kittelson & Pamela J. Transue (eds.), *Rebirth, Reform, and Resilience: Universities in Transition 1300–1700* (Columbus: Ohio State University Press, 1984), 77.

The shortage of books was one reason for the early pre-eminence of reading aloud. Since copies were rare and expensive, such methods of instruction served as important channels for knowledge transfer. However, the function of *lectio* apparently changed down the centuries. Originally, at least according to a couple of commentators, this practice was primarily associated with the significant oral elements of medieval culture.[8] Dictation provided a written record intended to preserve the living spoken word, which ultimately stemmed, not from the teacher, but from the ancient and biblical authorities and their interpreters. When the art of printing became widespread, this subsequently forced, as one commentator puts it, 'a clarification of the purpose of university studies'.[9] The function of transmission gradually altered, and the students seem to have increasingly come to understand and treat what was said as being derived from an authentic, invariable, and typographically fixed text.[10] Gutenberg's invention thus renewed the view of the lecture, even if actual tuition practices still harked back to older traditions. It should be added that, paradoxically, this older tradition lacked a developed theoretical reflection on, and discussion of, lecture practice. It was only during the eighteenth century that the lecture began to be regarded as a rhetorical genre in its own right.[11]

Oral teaching also had a different function during its early history. It made the monitoring of the university by external bodies possible, and it was common for lecturing to be strictly regulated in the statutes of the universities. Nonetheless, lecturing per se seems to have been appreciated: students gained access to a defined and fixed subject matter that was manageable; professors, for their part, could keep their preparations to a minimum. They had some room for manoeuvre with regard to sensitive topics and inappropriate references, which could not be discussed in dissertations. The authorities that regulated the activities of the universities, however, often closely watched how the lecturers performed their duties, and the students' attendance was often compulsory. Lectures were given daily, or at least several times a week. They were supplemented by other academic activities such as disputations. A lecture generally lasted an hour or so. Saturday, Sunday, and one weekday were usually

8 See Stichweh 1994, 289; and Grant 1984, 98 n. 47.

9 Stichweh 1994, 286.

10 On the genesis of such conceptions of the written word, see Thomas Götselius, *Själens medium: Skrift och subjekt i Nordeuropa kring 1500* (Gothenburg: Glänta, 2010), 123 *et passim*.

11 See Lindberg 2011, 40.

free. Often the lecture-halls were designed so that especially wealthy or famous students were given the best seats. Membership of student *nationes* was another means of segregating the students.

The Emergence of the Original Teacher

In the decades leading up to the turn of the nineteenth century, established teaching methods faced increasingly fierce opposition. Reform-minded educators and university debaters conducted a polemic against the established institutions, and their criticisms were a symptom of a general decline. Throughout Europe, the backward-looking and old-fashioned universities were challenged. New knowledge was increasingly produced outside the confines of the universities, whose aversion to renewal was also reflected in the fact that the number of enrolled students fell, especially in the German-speaking regions: this, the traditional form of higher education, was becoming unattractive, especially to young men from the elite group thought to be the pillars of society. This development is understandable for reasons including the drop in teaching quality at several universities, where soulless recitation from set texts and unintelligible comments were legion. What was the actual motivation behind lecturing from the podium, reproducing, without any insight or personal touch, what had long been established, while the students passively recorded this information? This issue had indeed already become evident as the art of printing developed. However, around the turn of the nineteenth century the German-speaking regions witnessed a widespread debate in which several different positions were represented.[12]

One of the solutions that emerged was for the lecture to be placed on a new footing; it was no longer thought reasonable to merely reproduce accepted wisdom fixed in text. In a culture where the medium of the written word was now decidedly pre-eminent, the *raison d'être* of the universities could not be the exclusive control of this medium. The one-way transmission of tradition by the reading of texts had no place here. Instead, a new idea began to exert its influence, to the effect that oral presentation must be a genuinely productive process, where the lecturer acted in an open and creative space. The lecture should not be based on set Latin texts, but was increasingly considered a unique process, which no book reading could replace, and where it was not legitimate merely to read aloud from a prepared script.

---

12 See Peter Josephson, 'Böcker eller universitet? Om ett tema i tysk utbildningspolitisk debatt kring 1800', *Lychnos: Årsbok för idé-och lärdomshistoria* (2009).

The notion that a teacher should be original now became commonplace. He was no longer a *lector* but an *auctor*, an author. The professor vouched for the veracity of what he (for a long time to come, it was always a he) transmitted. By contrast, he was not responsible in the same way as before for the correlation between the subject matter and established knowledge. Individual judgement and autonomous thought became the right and duty of the university teacher. In other words, the lectern was transformed into a central place for originality.

The task for the lecturer in this new age was thus not to pass on what had been handed down, but to embody the creation of new knowledge. In addition to his teaching duties, he had also to act as a researcher. Those who were to be capable of guiding others had themselves to produce new knowledge. It was here that the axiom was postulated as to the irrefutable connection between teaching and research that is nowadays associated with Humboldt's name. This also applied to the audience. According to the new ideal, the listening students should also reflect and act creatively in the lecture situation. An oral presentation before an audience was viewed as a dialogical process, whereby those who sat silently in the lecture-hall were also regarded as participants in a discourse.

One of the first teachers in the German-speaking area to act in accordance with these changed ideals was Johann Gottlieb Fichte. In the mid 1790s, while teaching philosophy in Jena, he partly began to lecture without notes or a set book.[13] At the same time, these new ideas also took hold at the reformed educational institutions that were opened in France during the Revolution; here, too, demands were made that university teachers should give free-form lectures and actively address the audience, although it was on German soil that these new approaches gained a conceptual legitimization.[14]

This was thanks largely to such individuals as Friedrich Schelling, the leading representative of Romantic philosophy, who was among the first to specifically highlight the idea that teaching and research must form a unity. He formulated his new ideas in a series of lectures on the objectives and meaning of university studies, which he delivered at Jena just after the turn of the nineteenth century.[15] The lecturer could no longer merely reproduce what had been gleaned from others; instead, he must be an original creator, even in the

13 See Adelheid Ehrlich, *Fichte als Redner* (Munich: TUDUV Verlagsgesellschaft, 1977), 26 *et passim*.

14 See Apel 1999, 60.

15 Friedrich Wilhelm Joseph Schelling, *On University Studies*, trans. E.S. Morgan (Athens: Ohio University Press, 1966).

midst of presenting his lecture, otherwise one could just as well refer the students directly to the printed texts—that is, to 'popular textbooks expressly written for him or to voluminous compilations that exist in every branch of knowledge'.[16] The teacher must therefore bring knowledge to life in front of his students and draw his audience into a genuine exchange that no text can replace. According to Schelling, nothing could be more 'mind-killing' than a passive approach to teaching.[17] Indeed, the very purpose of the academic lecture was to ensure the coming into being of what is being discussed. For the 'real advantage of live instruction', Schelling says, is that 'the lecturer does not merely give results'.[18] Instead, he shows 'how these results were reached, and at every point builds up the whole science, as it were, before the student's eyes'.[19]

The ideals formulated in Jena were soon to be brought to Berlin. In the discussions that preceded the founding by Wilhelm von Humboldt of the new seat of learning in the Prussian capital, several leading commentators advocated these new ideas. The theologian Friedrich Schleiermacher's well-known proposal *Gelegentliche Gedanken über Universitäten in deutschem Sinn* (1808) contains, for example, a number of characteristic formulations on the function of the podium lecture.[20] The function of oral teaching, according to Schleiermacher, must be to awaken an inner process of education in every individual. Therefore, this practice should refer back to antiquity and its maieutic methods rather than to the university's traditional, one-way transfer of knowledge through the reading of texts. 'The teacher', wrote Schleiermacher in a formulation that recalls Schelling, 'must let everything he knows arise in front of the audience'.[21] Hence, he could not merely recount, but should instead reproduce the genesis of 'his insight, the action itself'.[22] The lecturer, in other words, had to possess an enthusiastic and living knowledge of his subject, with which he could illuminate the educational process and avoid the soulless

16 Ibid. 26.

17 Ibid. 27.

18 Ibid. 27.

19 Ibid. 27.

20 Friedrich Daniel Schleiermacher, 'Gelegentliche Gedanken über Universitäten in deutschem Sinn', in Wilhelm Weischedel (ed.), *Idee und Wirklichkeit einer Universität: Dokumente zur Geschichte der Friedrich-Wilhelms-Universität zu Berlin* (Berlin: Walter de Gruyter, 1960), 136; see also, for example, Johann Gottlieb Fichte, 'Deduzierter Plan einer zu Berlin zu errichtenden höheren Lehranstalt', in Weischedel 1960, 30; Christoph Wilhelm Hufeland, 'Ideen über die neu zu errichtende Universität zu Berlin und ihre Verbindung mit der Akademie der Wissenschaften und anderen Instituten', in Weischedel 1960, 17.

21 Schleiermacher 1960, 137.

22 Ibid. 137.

repetition of a script. In this manner, he would give new meaning to the art of lecturing in an age of mass-produced printed texts. The students would become elements in the knowledge process, in that their spirit would be awakened to reflection and contemplation. Clearly, the audience itself was expected to remain silent during a lecture, but their thinking was nonetheless formed into a living dialogue with the lecturer.

When, about a year later, Humboldt commented on the academic lecture in his memorandum, he thus referred back to his predecessors. He had lived in Jena in the past, and had been deeply influenced by that city's intellectual milieu. Long before his Berlin period, he saw and heard young, radical philosophers lecturing to enraptured audiences. Even if he always cherished solitary and free study, he was also convinced of the importance of the dialogical community to the education process.[23] Lecturing thus resembled the discourse of the seminar, another feature of academic life that became central with the rise of the modern university.

One could consider the new ideas about lecturing espoused by Fichte, Schelling, Schleiermacher, and Humboldt as an apology for orality and a bulwark against the clear pre-eminence of printed culture. Certainly, this defence was composed at the very time when the distinction between oral presentation and the printed word had become far clearer than had previously been the case.[24] Indeed, the fact is that the Romantics' rush to the lectern is clear evidence of the media historical insight that different modes of transmission must be conceived of as having a mutual relationship; modes that are often part of an interaction where various media impinge on one another. When printed literacy spread, it became necessary to imbue the spoken word with new meaning and its practice—in the form of academic lectures—with new legitimacy. This process, moreover, had clear parallels with other efforts to preserve the centrality and cultural status of the spoken word, as the culture of the printed word spread.[25]

Criticism of the Lecture

The Romantic and Humboldtian idea of the lecture has been constantly questioned over the past two centuries. As early as the end of the eighteenth

23 See Thomas Karlsohn, 'Humboldt och skriften. Om bildningens mediala förutsättningar', *Lychnos: Årsbok för idé- och lärdomshistoria* (2011), 219.

24 Cf. McDowell 2010.

25 Cf. ibid. 241.

century, those eager for a reform of university education were well aware that the happy combination of eminent researcher and skilled teacher is rare. Throughout the nineteenth century, various forms of criticism were heard, apparently prompted by the major shortcomings in teaching performance—frequently with the dialogic seminar played off against the monologue from the lectern.

The complaints continued throughout the nineteenth century and into the twentieth, with commentators disparaging lectures as antiquated relics of the medieval university, and protesting that the idealized manifesto of the Romantics had no counterpart in the real world. Change, however, would not come until the post-war period, especially in connection with the radicalization of the 1960s and 1970s. Suspicions grew on the part of educational researchers and keen reformers of educational policy in the German-speaking area and elsewhere; however, the politically tinged accords with the universities also resulted in bitter criticism from students of the lecture as an institution. These anti-authoritarian ideas amounted to a polemic against inherited practices, and traditional teaching methods were attacked because they impeded democratic influence. The students questioned the uncritical adoption of older teaching methods, and held that they as listeners had little opportunity to argue against the truths that were being presented.

Another objection to traditional oral authority was directed at the tendency to fall back on seductive rhetoric and acting tricks at the expense of content and a meaningful acquisition of knowledge. For example, what became known as the Dr Fox effect was discussed in the 1970s. This took its name from a certain Dr Myron L. Fox, who travelled around the US lecturing on medicine. His appearances were much talked about, he received excellent reviews from the students he met, and his originality seemed obvious. However, in reality Fox was an actor and completely lacked any profound insight into the subjects he taught. His activities were instead part of a research project that examined the harmful consequences of charismatic audience appeal.[26]

Weighty objections to the lecture as a form of teaching were also articulated in the theories of learning that were gaining currency during the same period.[27] It was widely felt that an oral presentation to an academic audience or a school

26 See Donald H. Naftulin, John E. Ware Jr, & Frank A. Donnelly, 'The Doctor Fox Lecture: A Paradigm of Educational Seduction', *Journal of Medical Education*, 48 (1973); John E. Ware Jr & Reed G. Williams, 'The Dr Fox Effect: A Study of Lecturer Effectiveness and Ratings of Instruction', *Journal of Medical Education*, 50 (1975).

27 Aspects of the discussion in the Anglo-Saxon context are summarized in Geoff Isaacs, 'Lecturing Practices and Note-Taking Purposes', *Studies in Higher Education*, 19 (1994), 203.

class was a one-sided and context-free transmission of the subject matter, and was in an unfortunate way detached from the concrete situations in which such knowledge could be used. One example of such objections is found in the emergence of the later widespread problem-based learning (PBL). The ideas underlying this new didactic concept emerged from experience of the obvious shortcomings of the older methods of teaching. Particular criticism was directed at the lecture-based transmission of facts in medical education. To counterbalance this, there was a desire to enhance the practice-related and clinical problem-solving closely linked to students' future professions. Oral presentations before a vast audience were replaced by active work in groups with the support of supervisors.

PBL is a singular, albeit influential, example taken from the multifaceted theoretical and practical development of education in recent decades. The overall trends have undoubtedly constituted a move away from lecturing. Constructivist approaches and the increasing emphasis on self-directed activities have contributed to a suspicion of lectures. The influential socio-cultural theory, with its focus on the situational and collaborative acquisition of knowledge, puts the emphasis elsewhere than on the traditional lecture. Admittedly, these theoretical trends have not affected university education in the same way that they have the school system;[28] however, they have played a decided role in the higher-education debate.

The Lecture Today and in the Future

The critique of oral teaching over the past two hundred years, and especially during the post-Second World War period, could be interpreted as a long, drawn-out questioning of the Romantic legacy. According to their adversaries, the ideas and practices of Humboldt and his contemporaries were flawed, because lecturing cannot achieve anything of real significance in the listeners' learning processes. Traditional oral presentations to an audience render the students passive and incapable of harnessing all their dormant, innate potential. Even in our times, such repudiation is commonplace.[29]

28 Cf. Diana Laurillard, *Rethinking University Teaching: A Conversational Framework for the Effective Use of Learning Technologies* (London: Routledge, 2002), 12.

29 See, for example, Stephen Sheely, 'Persistent Technologies: Why Can't We Stop Lecturing Online?', *Proceedings of the 23rd Annual Ascilite Conference: Who's Learning? Whose Technology?* (Sydney: University of Sydney, 2006), 769; Laurillard 2002, 93 *et passim*. The literature treating the problem of the lecture today on a general level is sparse. Valuable

If these reproaches had been justified in all respects, one might assume that lecturing would have ceased long ago. Yet this is not the case. The traditional lecture to a live audience still enjoys a strong position in many places as one of the more common didactic practices in higher education. Its popularity among students is often high, even though the vast majority of university teachers are unable to fully live up to established ideals. It is common to hear calls for more lectures, not fewer.

It has been argued that the survival of the lecture is due to educational slackness, ingrained habits, and set ideas about learning as a simple transfer of information from a sender to a recipient.[30] However, I believe that the explanation lies more in the fact that the oral presentation to an audience can often engender something more than an uncertain transfer of a given content—in other words, the educational philosophy developed two centuries ago took account of an important aspect of the academic lecture. Admittedly, lecturing is not always as effective or easy a means of knowledge transmission as other methods.[31] But it has survived because it makes possible experiences that compensate for its disadvantages.

What are these redeeming features? A contemporary critic, Diana Laurillard, dismisses adherence to traditional oral teaching, and holds that university teachers stubbornly 'defend the value of the "inspirational" lecture, as though this could clinch the argument'.[32] She poses the polemical question of how many inspirational performances it is possible to give in a week, and how many a student can absorb. Such questions imply a common but overly narrow view of the conditions for the lecture and its effects. What Laurillard terms inspiration is depicted as the lecturer's property, which under optimal but rare conditions can be transferred to the people present in the lecture-hall. However, I believe that those who today defend the lecture often refer to something more fundamental. As their predecessors did at the turn of the nineteenth century, they point to more essential phenomena than the inspirational ability that, at worst, depends only on superficial acting techniques similar to those mastered by Dr Fox. I should like to briefly mention two of these phenomena before I conclude with some thoughts on the new trend of e-lecturing.

---

contributions include Dusini & Miklautsch 2007; Konrad Liessmann, *Über Nutzen und Nachteil des Vorlesens: Eine Vorlesung über die Vorlesung* (Vienna: Picus, 1994); and Eberhard Straub, 'Vom Knattermimen zum Talkmeister', *Wirtschaft & Wissenschaft*, 4 (2007).

30 See, for example, Sheely 2006, 770; and cf. Laurillard 2002, 93.

31 See Donald A. Bligh, *What's the Use of Lectures?* (San Francisco: Jossey-Bass, 2000), 8.

32 Laurillard 2002, 93; cf. Bligh 2000, 14 & 59.

One important experience that can be gained from lecturing to an audience is uncertainty. The individual interaction with printed texts and other tuition materials is under fortunate circumstances supplemented by an unpredictable public process in the lecture-hall. During a lecture, there is always the risk of unexpected events. If the speaker is sensitive to the situation, the feeling of a dialogical exchange may become palpable; the lecture, much like the seminar, can then stand out as something distinctive. Both lecturers and listeners feel that the subject matter addressed does not equate with pre-packaged information to be conveyed or acquired in the simplest possible way; instead, knowledge appears as something changeable that is to be assimilated in an interpersonal exchange, and coloured by personal circumstances. This impression may very well be conveyed even if the lecturer is not particularly inspiring in a way that wins over the audience. Furthermore, the experience of uncertainty is not unlike that which characterizes a theatrical performance, concert, or football match. No written description, no recording can replace being physically present in the auditorium.

A similar issue is that of embodiment. The physical appearance of the lecturer and the place can together convey the fact that everyone knows intellectually, but has not always incorporated emotionally—that knowledge does not hover without context in a vacuum, devoid of attachment. Even if it is perceived as uninteresting, the lecture still amounts to a confirmation that the subject matter has real significance in the same world as the one that the audience inhabits. The event in the lecture-hall implies that the audience, through the medium of a real person, become tangibly acquainted with research. The lecturer offers insight into and experience of an environment that represents values and attitudes within a large academic community of which they themselves are also a part.

These two experiences, of uncertainty and embodiment, appear to be threatened in the verbal instruction format that has become popular in the most recent phase in the history of the lecture. At many seats of learning around the world, fibre-optically mediated e-lectures have thus swiftly come to replace traditional teaching.[33] Oral performances are streamed live or videoed and posted online. Quite frequently, the same lecture is reused several times, which allows for cost savings. Especially in countries where higher education is under considerable commercial pressure, e-lectures have become common. In some contexts, there is a clear stratification, where financially secure

33 For a comprehensive and critical discussion of e-lecturing, see Tara Brabazon, *The University of Google: Education in the (Post) Information Age* (Aldershot: Ashgate, 2007), 103.

students can afford courses with live instruction, leaving the less well-off to make do with YouTube.

The e-lecture has historical antecedents, including experiments with televised teaching in the 1960s and 1970s.[34] However, because of the spread and ease of use of digital technology, oral teaching transmitted via visual media has become something to be taken seriously. At the same time, this is a paradoxical trend. The increasing activity in this field is in stark contrast to many influential educational ideas that present human learning as interactive and context-bound. Admittedly, its advocates have already asserted that the traditional lecture is a problematic method, which conveys a false impression of 'learning in manageable, discrete, self contained bundles'.[35] However, as other commentators have observed, rather it is the e-lecture that decontextualizes learning and presents a picture of knowledge as something easily consumable.[36]

Despite these objections, many people are enthusiastic. Not that the reason for their enthusiasm is the evident, unambiguous benefits. There is, for example, no evidence that e-lectures enable learning that is more effective.[37] Instead, financial interests are often sensed in the underlying context, even if honest attempts at a didactic improvement obviously also elicit a willingness to try new solutions based on digital technology. However, one factor in all this is the fact that students still seem to appreciate the lecture-hall.[38] Studies show

34 A Swedish example is discussed in detail in Lennart Sturesson, *TV som undervisningsteknologi: Examplet Linköpings tekniska högskola* (Lund: Arkiv, 2005). There are analogies in many other countries, not the least in North America. See for instance Larry Cuban, *Oversold and Underused: Computers in the Classroom* (Cambridge: Harvard University Press, 2002), 99.

35 Sheely 2006, 772.

36 Emily Bennett & Nipan Maniar, 'Are Videoed Lectures an Effective Teaching Tool?', <http://podcastingforpp.pbworks.com/f/Bennett+plymouth.pdf>, accessed 5 October 2010.

37 See Ashley Deal, 'Lecture Webcasting: A Teaching with Technology White Paper', in id., *Research on Teaching with Technology* (Pittsburg: Carnegie Mellon University, 2007), 5; Stavros Demetriadis & Andreas Pombortsis, 'E-lectures for Flexible Learning: A Study on Their Learning Efficiency', *Journal of Educational Technology & Society*, 10 (2007).

38 See, for example, Troy Cooper, 'Reports of the "Death of Geography" Have Been Greatly Exaggerated: Why UK Distance Learning Students Prefer Face-to-Face Tuition', in Ulrich Bernath, András Szücs, Alan Tait & Martine Vidal (eds.) *Distance and E-Learning in Transition: Learning Innovation, Technology, and Social Challenges* (London: ISTE, 2009), 543; Jeffrey R. Young, 'The Lectures Are Recorded, So Why Go to Class?', *Chronicle of Higher Education*, 36 (2008); but cf. Scott Cardall, Edward Krupat & Michael Ulrich, 'Live Lecture Versus Video-Recorded Lecture: Are Students Voting With Their Feet?', *Academic Medicine*, 12 (2008), 1174.

that if they have access to the same lecture as a recording or in a lecture-hall, the proportion of listeners present still does not decline.[39] The e-lecture is regarded as a complement rather than as a replacement. As such, it can of course function very well, although switching between different media formats can be problematic.[40] At the same time, the students' reactions show that traditional oral teaching still has a role to play. These reactions are probably due to the distinctive experiences made possible in the auditorium. The historic legacy of Humboldt's insight that a performance before an audience can be something more than a simple transfer of knowledge seems to live on. And therefore the auditorium ought still to be regarded as an important location at every college and university.

---

39 Deal 2007, 4; Simon Fietze, 'Podcast in Higher Education: Students Usage Behaviour', *Proceedings ascilite Auckland* (2009), 314 *et passim*; and cf. David Hearnshaw, 'Will Podcasting Finally Kill the Lecture?', *The Guardian*, 19 September 2006.

40 See Jeffrey E. Young, 'Short and Sweet: Technology Shrinks the Lecture', *The Chronicle of Higher Education* (54) 2008.

CHAPTER 3

It Takes a Real Man to Show True Femininity

Gender Transgression in Goethe's and Humboldt's Concept of Bildung

Claudia Lindén

'All that is masculine displays more independence, all that is feminine more suffering receptivity', Humboldt wrote in 1795.[1] For Humboldt, genius, art, and *Bildung* followed a heterosexual structure with an active man and a passive woman. For Goethe, *Bildung* in women was not attractive, or as he bluntly put it, 'We love in a young woman things entirely different from her intelligence'.[2]

When we refer to the concept of *Bildung* today it is used as a gender-neutral ideal for university culture. Yet *Bildung,* as the quotations from Humboldt and Goethe show, was hardly gender-neutral in the eighteenth century when the concept was coined. At a time when women were not citizens and had no access to the university, all talk of *Bildung* silently implied men. Reinhart Koselleck points out that the German concept of *Bildung* concerns something larger and more amorphous than just education (and therefore theoretically could belong to women). Yet Koselleck confirms, in the same silently implied manner, *Bildung*'s masculine connotations when he, without any further reflection, lists all the possible ways in which *Bildung* was articulated. They turn out to be only fields where women could *not* partake until the late nineteenth century.[3] Not only is the presumed subject of *Bildung* male, but also notions of sexual difference lie at the very heart of the concept of *Bildung*. That does not mean that we cannot use *Bildung* today as a general human ideal, but we should be aware of its gender-differentiated, heteronormative, and misogynist history.

1 Wilhelm von Humboldt, 'Über Geschlechtsunterschied und dessen Einfluß auf die organishe Natur', *Schriften zur Antropologie und Gesichichte*, in *Werke in fünf Bänden*, ed. Andreas Flitner & Klaus Giel (Darmstadt: Wissenschaftl. Buchgesellschaft, 1960), 277 quoted in translation in Catriona MacLeod, *Embodying Ambiguity: Androgyny and Aesthetics from Winckelmann to Keller* (Detroit: Wayne State University Press, 1998), 48.

2 Johann Peter Eckermann, *Gespräche mit Goethe* (Leipzig: Brockhaus, 1909), 428 quoted in translation in James Albisetti, *Schooling German Girls and Women: Secondary and higher Education in the Nineteenth Century* (Princeton: PUP, 1988), 3.

3 Reinhart Koselleck, 'On the Anthropological and Semantic Structure of *Bildung*', in id., *The Practice of Conceptual History: Timing History, Spacing Concepts*, trans. Todd Presner, Kerstin Behnke & Jobst Welge (Stanford: SUP, 2002), 171 & 175.

 | DOI 10.1163/9789004271944_005

The Romantic period, especially in Germany, is known for its strong emphasis on the polarization of gender roles.[4] Yet at the same time it was the culmination of a period were gender and sex were not as clearly defined as they would be later in the nineteenth century. Literature at the end of the eighteenth century explored gender ambiguity, transgression, and queer desires of all kinds, performing heteronormative sexual difference all the while.[5] Feminist historians of the French Revolution have pointed out that the public sphere was constituted through the exclusion of women.[6] The exclusion of women was accompanied by a strong ethos of male friendship that created an internal division within masculinity that provided the possibility of both a 'masculinized' and a 'feminized' masculinity.[7] The art historian Abigail Solomon-Godeau has shown that the male nude in French neoclassical art in the late eighteenth century is represented as feminized, ephebic, and passive. Her point is that this is related to a political and cultural change in masculinity and femininity, which produced a representation of masculinity that contained femininity.

As a scholar working at the intersection between comparative literature and gender and queer theory, my interest in Goethe and Humboldt concerns how gender transgression, or a mixing of male and female, becomes an intrinsic part of masculinity in Goethe's *Wilhelm Meister* and Humboldt's 'On sexual difference' (1794) and 'On the male and female form' (1797). How do their views

4 See, for example, Christine Battersby, *Gender and Genius: Towards a Feminist Aesthetics.* (London: Women's Press, 1989); Kurt Lüthi, *Feminismus und Romantik: Sprache, Gesellschaft, Symbole, Religion* (Vienna: Böhlau, 1985); Anne Kostelanetz Mellor (ed.), *Romanticism and Feminism* (Bloomington: Indiana University Press, 1988); Friedrich A. Kittler, *Discourse Networks 1800/1900* (Stanford: SUP, 1990); Anne Kostelanetz Mellor, *Romanticism and Gender* (New York: Routledge, 1993); Elizabeth A. Fay, *A Feminist Introduction to Romanticism* (Malden, Mass.: Blackwell, 1998); Marc Redfield, *The Politics of Aesthetics: Nationalism, Gender, Romanticism* (Stanford: SUP, 2003).

5 There is a growing body of work on this period within the field of queer theory. See, for example, Alice A. Kuzniar (ed.), *Outing Goethe and his Age* (Stanford: SUP, 1996); Robert D. Tobin, *Warm Brothers: Queer Theory and the Age of Goethe* (Philadelphia: University of Pennsylvania Press, 2000); Peter Cryle & Lisa O'Connell (eds.), *Libertine Enlightenment* (Basingstoke: Palgrave Macmillan, 2003); Susan E. Gustafson, *Men Desiring Men: the Poetry of Same-Sex Identity and Desire in German Classicism* (Detroit: Wayne State University Press, 2002); and Max Fincher, *Queering Gothic in the Romantic Age: The Penetrating Eye* (Basingstoke: Palgrave Macmillan, 2007).

6 Joan W. Scott, *Only Paradoxes to Offer. French Feminists and the Rights of Man* (Cambridge, Mass.: Routledge, 1996), 10.

7 Abigail Solomon-Godeau, *Male Trouble: A Crisis in Representation* (London: Thames & Hudson, 1997), 13.

on gender relate to the contemporary and highly politically charged debates on gender that emanated from the demand for women's education and citizenship during the French Revolution? In this chapter I will argue, in line with Solomon-Godeau's idea of a particular representation of masculinity with an inner femininity, that Goethe and Humboldt's representation of masculinity can be read as part of the same political and aesthetic discourse that produced not only a new, biologically based misogyny, but also a new interest in a 'feminine' masculinity. My aim is not to discuss their views on women per se, but a new type of masculinity that they both contributed to creating. More specifically, I want to elucidate a particular sense of androgyny at work here, which appears to be a prerogative for *Bildung*, and yet somehow only accessible to men.

Goethe, Humboldt, *Bildung*, and Gender

Very little has been written specifically on *Bildung* and gender. A search in library catalogues gives only a few titles on women and education in Germany that mention Humboldt's two texts on gender from 1794 and 1795.[8] There is no study devoted to the concept of *Bildung* from a gender perspective in the way that has been done for the concepts of 'genius', 'the sublime', or 'the artist'.[9] On the other hand there is quite a bit of research on women and education from this time, especially from British and French perspectives. This is not a surprise, since the discussion of education for women is strongly connected to feminist thought (and to an anti-feminist stance that is manifest in many of the books on the proper upbringing of young women) and it is out of Britain and France that the early feminism grew at the time of, and partly in reaction to, the French Revolution.[10] Even though education is not

8 Albisetti 1988, 5 & 15; Peter Petschauer, *The Education of Women in Eighteenth Century Germany: New Directions From The German Female Perspective Bending the Ivy* (Lewiston, N.Y: E. Mellen Press, 1989), 278; Martha B. Helfer Ivy, 'Gender Studies and Romanticism', in Dennis F. Mahoney (ed.), *Camden House History of German Literature, vol. 8, The Literature of German Romanticism* (Rochester, N.Y: Camden House, 2004), 234.

9 For example, Battersby 1989; Barker Deborah, *Aesthetics and Gender in American Literature: Portraits of the Woman Artist* (Lewisberg, NJ: Bucknell University Press, 2000); or the classic essay by Linda Nochlin, 'Why are there no great women artists?', *Aesthetics* (1998), 314–323.

10 I will use the term 'feminism' to refer to authors, texts, and political events that openly strived to change women's situations and that were linked to the history of feminist thought.

the same as *Bildung*, the discussion of women, reason, and education at the time of Goethe and Humboldt is a valid context for analysing their views on masculinity, reason, and *Bildung*, and I will draw from this research when necessary.

Gender-oriented research in studies of Wilhelm von Humboldt is scarce. The only one who has written more extensively on Humboldt and these particular texts is Catriona McLeod in her book *Embodying Ambiguity. Androgyny and Aesthetics from Winkelmann to Keller* (1998). McLeod points out in her reading of Humboldt that his concept of androgyny is firmly grounded in heterosexual complementarity, as compared to Winkelmann's more hermaphroditic androgyny. McLeod's reading of Humboldt is useful for my argument, but since her focus is androgyny and not masculinity I will extend my reading of Humboldt to a larger context of gender formation.

The literature on Goethe, on the other hand, is abundant, and there are several studies on Goethe's view of women from a gender perspective, even though this is not a central focus in Goethe research.[11] Rather than feminist readings of Goethe, there is today a growing interest in Goethe from a perspective of queer studies. The many instances of gender transgression, homoerotic desire, and queer sexuality in Goethe's texts make him interesting for such readings.[12] Here again I build on what others have done before me, but as with Humboldt, my contribution lies in a focus on masculinity and in contextualizing it in terms of the discussions of women's rights and views on art at the time.

A problem when one wants to do research on historical gender with general concepts such as *Bildung*, citizenship, the sublime, and so on, is that when they were coined the notion of the 'human' was almost always synonymous with man. It was not even necessary to mention that the human subject of a specific treatise was an implied man, since it was just taken for granted. The scholar interested in gender constructions therefore often has to go to marginal and insignificant texts of famous authors to find what they had written specifically on sexual difference. Kant is a famous example: he does not write much on

11 See, for example, Eleanor E. ter Horst, *Lessing, Goethe, Kleist and the Transformation of Gender: From Hermapfrodite to Amazon* (New York: Lang, 2003); Benjamin Bennett, *Goethe as Woman: The Undoing of Literature* (Detroit: Wayne State University Press, 2001).

12 The Goethe scholar Richard Tobin gives an example of resistance to highlighting queer traits in German Romaticism when he talks about the angry reactions to attempted queer readings of Goethe. Richard Tobin, *Warm Brothers. Queer Theory and The Age of Goethe* (Philadelphia: University of Pennsylvania Press, 2000), x.

women at all, but in the third section of *Beobachtungen über das Gefühl des Schönen und Erhabenen* (1764; *Observations on the Feeling of the Beautiful and Sublime*, 1799) he offers an account of sexual difference that imagines men and women as complementary. But it is in *Grundlegung zur Metaphysik der Sitten* (1785; *Grounding for the Metaphysics of Morals*, 1895/2005) that it becomes clear that sexual difference in his view also comes with a hierarchy of values, or as Beverly Clack writes: 'Kant equates masculine values with the qualities necessary for genuine moral action'.[13] It is duty, and not sympathy and feeling, which is associated with women. Peter Petschauer points out that Kant, like many others, 'thought that women would lose their beauty if they entered the male sphere of the noble and the sublime; that they would internalize male traits by acquiring intellect (*Verstand*)'.[14] On the whole, he continues, the likes of Schiller, Goethe, and Kant 'thought of men when they spoke of the ideal human being and reserved only the essence of being a mother (*Muttergeist*) for women'.[15] This is a way of working through indirect interpretation within Kant's work. It is an interpretative strategy that makes it possible to elicit an author's view on gender difference, even though this is not an explicit theme in his work.

Humboldt made no reference to the formal education of girls beyond the elementary level in his plans for the Prussian school reform in early 1800. Nor did he do anything to create secondary schools for girls during his time as head of the education ministry in 1809–1810.[16] Read with the same interpretative strategy as mentioned above, Humboldt's 'On sexual difference' (1794) and 'On the male and female form' (1797) give a clearer picture of why Humboldt neglected girl's education, since these texts are outspoken on gender difference in a way that later texts, like the famous *On Language*, are not. The same applies to Goethe's gender-transgressing women, when one reads Goethe's celebration of male actors playing women's parts as a higher art form.

Through a reading of Goethe's and Humboldt's texts and in the context of other texts, paintings, and feminist discussions at this time, it is possible to show how their specific form of feminized masculinity was a product of the political and aesthetic discourses of their time.

---

13 Beverly Clack (ed.), *Misogyny in the Western Philosophical Tradition: A Reader* (Basingstoke: Macmillan, 1999), 145. Kristina Fjelkestam, *Det sublimas politik* (Gothenburg: Makadam 2010) makes use of the same method.

14 Petschauer 1989, 281.

15 Petschauer 1989, 283.

16 Albisetti 1988, 15.

Education, Reason, Citizenship, and Politics

How did *Bildung* become an amalgamation of a 'masculinized' and a 'feminized' masculinity, reserved exclusively for biological males? There are three main steps to explaining this development, the first being women's demands for education and the right to citizenship at the end of eighteenth century, with its implication that women too have reason. How did this demand affect politics, philosophy, and art? The second is views on sex and gender in relation to the theatre during eighteenth century, especially Goethe's admiration of male actors taking female roles. Theatre and citizenship are related here since both were public spaces. The third step is to show, through a close reading of Goethe's *Wilhelm Meister* and Humboldt's early texts on gender and *Bildung*, how the notion of androgyny and sexual difference, so central to German Romanticism, became a part of the masculinity required for *Bildung* that is produced in these texts.

As already noted, I would argue that from a gender perspective, it is necessary to view the concept of *Bildung* in the context of early 'feminism' and the debate on women's education (by 1800, a debate that had been going on for a century).[17] During the latter half of the eighteenth century, the question of women's education became intrinsically linked to the question of women's citizenship and early feminism. Perhaps as a reaction to the 'feminist' books on education, or even concurring with them, this period saw a rapid growth of treatises such as Joachim Heinrich Campe's *Väterlicher Rath für meine Tochter* ('Fatherly Advice for My Daughter'), especially in German Protestant circles.[18] Peter Petschauer has pointed out that in the last twenty-five years of the eighteenth century there was an outpouring in Germany of studies on the female sex following 'the realization among the many persons writing and reading about it that women's education is intimately connected with women's role in society. Men and women now struggled over this underlying issue'.[19]

---

17 Before Mary Wollstonecraft took up the argument for women's rights, there were such texts as Mary Astell's tract on women's education, *A Serious Proposal to the Ladies* (1694), and Catherine Macaulay's *Letters on Education* (1790). Women took part in the discussion on women and education, promoting the idea that British and French women too had reason and should be educated. See Michèle Cohen '"To think, to compare, to combine, to methodise": Girls' Education in Enlightenment Britain', in Sarah Knott & Barbara Taylor (eds.), *Women, Gender and Enlightenment* (Basingstoke: Palgrave Macmillan, 2005), 224–242; and Jean Bloche, 'Discourses of Female Education in the Writings of Eigtheenth-Century French Women', in Knott & Taylor 2005, 243–258.

18 Albisetti 1988, 9.

19 Petschauer 1989, 283.

The most famous of these books on upbringing and conduct is, of course, Rousseau's *Émile ou De l'éducation* (1762), with its strong emphasis on sexual differences. In *Émile,* Rousseau outlines the boy's education first and then the girl's. But the girl is educated solely to be a companion to the man: 'It is not good for man to be alone. Emilie is a man. We have promised him a companion. She has to be given to him', says Rousseau.[20] The purpose of Sophie's education is to make her as pleasing as possible to Émile: 'Once this principle is established, it follows that the woman is made specially to please man. If a man ought to please her in turn, it is due to a less direct necessity. His merit is in his power; he pleases by the sole fact of his strength. This is not the law of love, I agree. But it is that of nature prior to love itself'.[21] The picture of the woman as someone whose main purpose is to help the man in his development was to return in several of the German *Bildung* texts, including Goethe's *Wilhelm Meister* and Schlegel's *Lucinde.*

The idea that women's role in marriage is to cater to the man's pleasure and needs was one of the things that Mary Wollstonecraft found particularly annoying about Rousseau. Several women, for example Mme de Staël, wrote articles against Rousseau's view of women and education. In Sweden, Hedvig Charlotta Nordenflycht wrote her famous poem *Till Fruentimmrens försvar* ('In defence of women') as a direct reaction to Rousseau's views on women in *Lettre d'Alembert.*[22] These texts show us that the Rousseauan ideal of womanhood as existing to help and inspire man in his education and *Bildung,* had already been launched *and* refuted by contemporary women by the time Goethe presented it in *Wilhelm Meister* or when Humboldt coined his androgynous ideal of *Bildung* for men.

One important reason why a celebrated eighteenth-century historian such as Catherine Macaulay could view the French Revolution as a pivotal historical moment was its promise that women too could become fully enfranchized citizens.[23] As we know, this was not to be—and the historical irony is that French

20 Jean Jaques Rousseau, *Emile: Or On Education,* translation Allan Bloom (New York: Basic Books, 1979) 357 and at http://archive.org/stream/RousseausemileOrOnEducationbloom/Rousseau-Emile, 363.

21 Rousseau 1979, 358 and at http://archive.org/stream/RousseausemileOrOnEducationbloom/Rousseau-Emile, 364.

22 A translation into French was planned, but Nordenflycht died before it could be completed. See Ann Öhrberg, *Vittra fruntimmer: författarroll och retorik hos frihetstidens kvinnliga författare* (Stockholm: Gidlunds, 2001).

23 Catherine Macaulay, *Observations on the Reflections of the Right Honourable Mr Edmund Burke on the Revolution in France* (London, 1790), and Mary Wollstonecraft,

women were among those who had to wait longest, until 1944—but for a brief period of time it looked as if a major historical and political change in society's division of sex and gender was to take place. Sex, as Thomas Laqueur puts it, 'was a major battleground of the French Revolution'.[24] And the whole world watched the events in France. If one is interested in the history of gender formation, the events of the French Revolution are crucial in a way that almost no other historical moment has been. It is therefore likely that both Goethe and Humboldt (who was in France just after the Revolution) would somehow react to this in texts written at the same time as the first demands for women's rights were heard in 1793. Friedrich Schiller most certainly did when he described the female revolutionaries: 'Women turn into hyenas' and 'with the teeth of panthers, / They tear out the hearts of their enemies', he wrote in his nightmarish poetic vision of the French Revolution, 'The Song of the Bell'.[25] Hyenas were believed to be truly hermaphroditic, possessing both female and male genitals and changing sex every year, and were used as a pejorative terms for feminists, as when Horace Walpole referred to Wollstonecraft as a 'hyena in petticoats'.[26]

As the French Constitution was debated in 1791, Olympe de Gouges published her *Déclaration des droits de la femme et de la citoyenne* ('Declaration of the rights of woman and the female citizen'). It was document that insisted that women by nature had all the rights of men, that they too were individuals, and that their specific needs as women made the exercise of those rights all the more urgent.[27] De Gouge's was not the first or only feminist statement on the Revolution, but it was, as Joan Scott writes, 'the most comprehensive call for women's rights in this period'.[28] The reaction against women's demands for equal rights was to underline sexual difference, and to situate femininity more firmly in the body.

Vindication of the Rights of Man, in a letter to the Right Honourable Edmund Burke (London, 1791).

24 Thomas Laqueur, *Making Sex: Body and Gender from The Greeks to Freud* (Cambridge, Mass.: Harvard University Press, 1990), 194.

25 Friedrich Shiller, *Gessammelte Werke*, i. 237 quoted in translation in Catriona McLeod, *Embodying Ambiguity: Androgyny and Aesthetics from Winckelmann to Keller* (Detroit: Wayne State University Press, 1998), 66.

26 Barbara Taylor, 'Introduction', in Mary Wollstonecraft, *A Vindication of the Rights of Women* (London: Everyman's Library 1992) xiii.

27 Olympe de Gouges, 'Declaration des droits de la femme et la citoyenne', *Ecrits politiques 1788–1791*, (Paris: Côté-Femmes, 1993) 204–215.

28 Scott 1996, 42.

Thomas Laqueur has shown in his celebrated book *Making Sex* that a fundamental shift in the way sexual difference was perceived took place during the latter part of eighteenth century. It is a shift from what he calls the one-sex model to the two-sex model. In the one-sex model, the sexes were the same in different degrees of metaphysical perfection, and the vagina was just an inverted mirror image of the penis. This model prevailed from antiquity until the end of the eighteenth century, which saw a change to the two-sex model, with fundamentally different sexes. 'By around 1800, writers of all sorts were determined to base what they insisted were fundamental differences between the male and the female sexes'.[29] Laqueur's point is that this change was not based on scientific discovery, but on political ideology; a change in language in the description of the sexes. He writes:

> The context for the articulation of two incommensurable sexes was, however, neither a theory of knowledge nor advances in scientific knowledge. The context was politics. There were endless new struggles for power and position in the enormously enlarged public sphere of the eighteenth century and particularly post-revolutionary nineteenth centuries: between and among men and women; between and among feminists and antifeminists. When, for many reasons, a pre-existing transcendental order or time-immemorial custom became a less and less plausible justification for social relations, the battleground of gender roles shifted to nature, to biological sex.[30]

In establishing a social and political distinction between men and women, genital differences made all the difference. According to this scheme, women were more profoundly defined by their sex than men. Many used Rousseau's famous words from *Émile*, where he states that since women have their genitals on the inside, as compared to men who have them on the outside, this 'internal influence continually recalls women to their sex'. This means, according to Rousseau, that 'the male is male only at certain moments, but the female is female throughout her life'.[31]

Scott writes that when 'women were definitely excluded from politics, their bodies were represented with obsessive frequency, most typically as nursing mothers'.[32] One way of establishing sexual differences at the time was, as the

---

29 Laqueur 1990, 5.

30 Laqueur 1990, 152.

31 Rousseau quoted in translation in Scott 1996, 49.

32 Scott 1996, 50.

feminist philosopher Geneviève Fraisse writes, to set up men and women in opposition to each other as good and evil.[33] Women were the dangerous others; the uncontrollable element in the flow of history; responsible for revolutions, and then guilty of derailing them. Such was the political, cultural, and aesthetic context when Goethe composed *Wilhelm Meister* and Humboldt his texts on gender from 1794 and 1795.

Bildung, Androgyny, Ephebic Men, and Sexual Difference

These examples of the rhetoric of women's suffrage at the time of the French Revolution indicate the close similarities between the political and the philosophical discourse. When women demanded to be part of reason, they were rejected, with reference made to their bodies—a move that secured the maleness of reason and citizenship even more strongly. Abigail Solomon-Godeau also makes the connection between male friendship and the new public sphere, where the new alliances between bourgeois men effectively required the exclusion of women.[34] The feminine returned to art in the shape of the ephebic male:

> In this reading, the ubiquity of the androgynous ephebe, exemplified by Meynier's *Eros* or Girodet's *Endymion,* is fostered not only by the homosocial or even homoerotic tenor of Neoclassical (elite) culture, but by the very cultural and political discourses concerned to expel (or contain) sexual difference itself. Such insistently feminized bodies can thus be interpreted as a return of the repressed, such that the feminine returns like a symptom, covertly inscribed, upon or within the body of the ideal youth.[35]

Solomon-Godeau's examples of representations of ephebic men in art at the end of the eighteenth century mirror the interest in androgyny, in hermaphrodites, Amazons, masculine women, and feminine men, that was so prevalent in German aesthetics and literature just at the time sexual difference was being converted into the two-sex model.[36] It is as if gender-bending or androgyny in

---

33 Geneviève Fraisse, *Reasons Muse: Sexual Difference and the Birth of Democracy,* trans. Jane Todd (Chicago: University of Chicago Press, 1994), 133.

34 Solomon-Godeau 1997, 11.

35 Solomon-Godeau 1997, 122.

36 Ter Horst 2003, 1; McLeod 1998, x.

one way or another emerged as an aesthetic response to the political and cultural gender turmoil of the day.

The concept of androgyny dates back at least to Plato's *Symposion*. Aristophanes tells the story of the different halves that once belonged together, and which were then split in two, forever desiring their original union. Androgyny as expressed in the Romantic aesthetic represented the desire for a fusing of dualistic structures. However, extreme antitheses do not necessarily cancel each other out in the middle; rather, they both take the middle ground and are preserved there intact. Many feminists have pointed out that the notion of androgyny, perceived as two separate entities meeting in the middle, still rests firmly on a binary ground, and that it continues to keep these binarisms intact. Catriona MacLeod goes on to connect this to a notion of the heteronormative: 'The key shift in eighteenth-century literary treatments of androgyny is I believe, that from a truly polymorphous bisexual model to an ideal of androgyny that regards heterosexual coupling as the norm and thus serves as the mythical cornerstone of an ideology of difference'.[37] So androgyny can work to preserve gender division, but in a new form, and as MacLeod continues, 'This harmonizing of binaries, with its promise of growth, is, of course, one of the fundamental concepts of classical *Bildung*'.[38]

In the same way as the ephebic man of Neoclassical art represents a way of containing sexual difference, my examples of Goethe and Humboldt will show two ways in which androgyny can function in the context of *Bildung*, stemming from an ideology of sexual difference and leading to *Bildung* being marked out as male. In both Goethe's and Humboldt's vision of androgyny, it is not as a fusion of opposites, but as a combination of binarisms of sexual difference. As in the pictures of Endymion or Eros, it is also a way of doing without women through a containment of sexual difference.

Gender Transgression, Masculinity, and *Bildung* in *Wilhelm Meister*

Goethe wrote *Wilhelm Meisters Lehrjahre* (*Wilhelm Meister's Apprenticeship*) in 1796, a book that was to be labelled the first *Bildungsroman*. Decades later he wrote the sequel, *Wilhelm Meisters Wanderjahre* (*Wilhelm Meister's Journeyman Years*). In the first Wilhelm Meister, male Bildung is created by resisting the temptations of sexual and erotic ambiguity. The novel ends with a male wholeness established, based on firmly defined gender roles within heterosexual

37 MacLeod 1998, 22.

38 MacLeod 1998, 42.

marriage, where man plays the role of reason and woman his helpmeet, just as Rousseau had prescribed.

The plot of the 600-page *Wilhelm Meister* is the story of how the eponymous hero is drawn to the world of theatre and drama even as a child.[39] The story opens with a meeting with his beloved Marianne, an actress. She is in love with Wilhelm, and pregnant with his child. But as an actress she is poor, and has to take other lovers to support herself. Wilhelm asks her to marry him, and she gladly accepts. They plan to start a theatre company together. But before the wedding can take place the other lover shows up, and Wilhelm, who accidentally sees the man, believes Marianne has deceived him. He is devastated, but never tries to talk with Marianne again or to contact her, despite the fact that he knows that she is expecting his child.

From the very first page of *Wilhelm Meister* the reader is thrown into a perplexing world of sexual ambiguity. Goethe presents a huge supporting cast of women bearing all the hallmarks of transvestism. The book opens with Marianne dressed in an army officer's uniform. But even though this has been part of her role, she wants to keep it when she is going to see her lover Wilhelm. Her male costume is depicted as having gone to her head, as she believes she can lead a life that transgresses the borders of traditional female life. Her lover, Norberg, wants her to be dressed as a woman in white so 'he can hold a little white lamb' in his arms.[40] Wilhelm, on the other hand, literally embraces the theatrical world of costume, travesty, and sexual ambivalence. When he meets Marianne and embraces her, it is the male costume, not the woman in it, that he holds close: 'How passionately didn't he embrace that red uniform and the white satin vest against his bosom!'[41] It is obvious that it is the fact that she is in men's clothing that makes her so attractive to him.

It is a known fact that Goethe showed a strong interest in the tradition of theatre where men or boys performed the female roles. This tradition had disappeared from France and Spain as early as the sixteenth century, and in Britain in 1660, but it was kept alive for another 150 years or so in Rome, Düsseldorf, and Vienna. Since this tradition was still alive in Rome at the end of the eighteenth century, where women were forbidden to appear on stage, both Goethe and Casanova have left records of their experiences of seeing male-only performances. Casanova wrote in 1762, having been to the opera to hear Giovanni Ost:

---

39 Johann Wolfgang von Goethe, *Goethe: The Collected Works, ix: Wilhelm Meister's Apprenticeship*, ed. Eric A. Blackall with Victor Lange (Princeton: PUP, 1995).

40 Goethe 1995, 40.

41 Goethe 1995, 2.

> Tightly laced, in a very well fitting corset, he had the figure of a nymph, and few women could show a firmer and more enticing bosom than his. The illusion he created was such that it was impossible to resist it. One looked, the spell acted, and one had either to fall in love or be the most stolid of all Germans...It was obvious that, as a man, he meant to foster love of those who loved him as such and who would not have loved him if he had not been a man; but he also meant to inspire love in those who, to love him, had to think of him as really a woman.[42]

Clearly, there was a pleasurable bisexual ambiguity about these men in women's clothes; it is obvious from Casanova's reflections that they had a strong allure for the audience. Goethe was a leading advocate of castrati in female roles. After the trip to Italy, he wrote an essay called 'Women's roles in the Roman theatre played by men'.[43] During his stay in Italy in January 1787 he had the chance to see an all-male performance of Goldoni's *La Locandiera* (1753), and this led him to suggest the replacement of actresses by actors across the board. For Goethe, a man performing femininity is more of an artist:

> I reflected on the reasons why these singers pleased so greatly, and I think I have found it. In these representations, the concept of imitation and art was invariably more strongly felt, and through their able performance a sort of conscious illusion was produced. Thus double pleasure is given in that these persons are not women, but only represent women. The young man has studied the properties of the female sex in essence and behaviour; he knows them thoroughly and reproduces them like an artist; he performs not himself but a third nature absolutely foreign to him. We come to know this nature even better because someone else has observed it, reflected on it, and presents us not with the thing itself but with the result of the thing.[44]

It was the art of mimesis, the imitation of life, that attracted Goethe.[45] It is only a performance as such when a man acts the woman's part. A woman doing the

42 Giacomo Casanova, *History of My Life*, vii. ch. 11, available at <www.gutenberg.org/>, accessed 1 December 2012.

43 'Frauenrollen auf dem römishen Theater durch Männer gespielt', *Gesammelte Werke* xiv. 9–13.

44 Quoted in Laurence Senelick, *The Changing Room: Sex, Drag and Theatre* (London: Routledge, 2000), 196.

45 It is interesting to compare Goethe's disapproval of women on stage with Rousseau's disapproval on moral grounds. Rousseau did not share Goethe's high regard for the

same would simply just *be* a woman; a mere display of nature, not an imitation of it. It becomes *art* only when a man impersonates a woman. To Goethe, men were better at being women. Goethe's views on castrati echoes those on the female men in French neoclassical art, and the cultural and political discourses concerned with expelling (or containing) sexual differences.

Eleanor Ter Horst rightly points out that Goethe's pleasure in these transvestite performances revealed distaste for the idea that there is truth in gender, and 'a discomfort with the two-sex model'.[46] Yet, she continues, it is only femininity that is revealed as artifice, and neither men who cross-dress as women nor women who cross-dress as men have any destabilizing implications for masculinity in *Wilhelm Meister*.[47] What Horst fails to see is that all the transvestite women, and with them all hints at same-sex desire, have to disappear from the story in order for Wilhelm to complete his *Bildung*. Masculinity is only stable when everything that threatens it has gone. Masculinity can, and should, take on femininity—witness the male actor playing a female part. It was the opposite—women acting like men—that should not happen.

Franco Moretti has stated that the *Bildungsroman* was about normality: 'It has accustomed us to looking at normality *from within* rather than from the stance of its exceptions, and it has produced a phenomenology that makes normality interesting and meaningful *as* normality'.[48] But something has to be expelled for normality to take place, and in the case of *Wilhelm Meister* that is all the androgynous women and, more importantly, the implied homoerotic desire linked to them. Robert Tobin has pointed out that Goethe's interest in same-sex desire, which can be found in several of his texts, always 'points to heterosexuality' in the end.[49]

---

theatre—something he expressed very clearly in his attack on actresses in *Lettre d'Alambert*. The actress becomes immoral just by being on the stage, for acting in itself is the art of illusion and pleasing, and a woman who professionally practiced such an art could surely be nothing but a prostitute. The bottom line for Rousseau was that women did not belong in the public sphere. Rousseau's and Goethe's engagement in keeping women off the stage must be viewed in relation to a broader interest in keeping women out of the public sphere. The question of women's citizenship lies at the heart of the discussion about women on stage as well as in *Bildung*.

46 Ter Horst 2003, 281.

47 Ter Horst 2003. 82.

48 Franco Moretti, *The Way of the World: The* Bildungsroman *in European Culture* (London: Verso, 1987), 11.

49 Robert D. Tobin, 'In and Against Nature', in Alice A. Kuznar (ed.), *Outing Goethe and His Age* (Stanford: SUP, 1996), 109.

Even if Wilhelm is attracted to Marianne because her transvestism, Goethe's disapproval of female actresses resonates through *Wilhelm Meister* in a series of portraits of bad actresses, most of whom die before the end of the novel. Marianne, Aurelia, and Mignon die, and the Baroness who likes to appear disguised as a 'page boy or a huntsman' only belongs for a while to the company.[50] When Wilhelm meets Mignon he does not know whether he is looking at a boy or a girl: 'He looked at the figure with amazement, uncertain whether it was a boy or a girl. But he finally decided in favour of the latter'.[51] The uncertainty of Mignon's gender and age, and of Wilhelm's relation to her, opens up to both homoerotic as well as incestuous desires. He wants to be like a father to her, but is not sure if the woman he had sex with in the dark was Mignon or not. When Wilhelm tells Mignon that he is going to marry Nathalie, she collapses and literally dies of a broken heart. Mignon's death thus confirms the undercurrents of same-sex and incestuous desire, since her death works to finally expel the last queerness from the story and to secure heteronormativity.

Franco Moretti argues that marriage is a central part of the plot in the *Bildungsroman*:

> the classical *Bildungsroman* 'must' always conclude with marriages.... Marriage as a metaphor for the social contract: this is so true that the classical *Bildungsroman* does not contrast marriage with celibacy, as would after all be logical, but with death (Goethe) or 'disgrace' (Austen).[52]

Marriage is the metaphor for the protagonists' relationship to society. One either marries or must, in one way or another, leave social life. According to this scheme, women who do not marry must die. As Peter Petschauer has pointed out, long before *Émile* was translated there existed in Germany a '*Hausväterliteratur*' that spoke of women's destiny as housewives.[53] As already mentioned, women were destined to be mothers, not to become *gebildet*. The androgynous women in *Wilhelm Meister* are important in driving the plot, but the female gender transgression Goethe describes there is something completely different from the type he was quick to praise in the male castrato. In a man, gender transgression is art; in a woman, it is the wrong kind of

50 Goethe 1995, 110.

51 Goethe 1995, 50.

52 Moretti 1987, 22f.

53 Petschauer 1989, 241 mentions Hohberg as an example of a book that was reprinted over and over again from late seventeenth and long into the eighteenth century.

transgression, one that has to be expelled from the story in order for the male hero to complete his *Bildung*.

Humboldt and *Bildung* as a Difference within

If gender-transgression in *Wilhelm Meister* finally has to be expelled from the story so that Wilhelm can ultimately secure his masculinity in a traditional heterosexual marriage with a woman who will be the female supporter for his continued *Bildung*, it is the other way around in Humboldt's vision of androgyny. For Humboldt, androgyny and femininity exists *in* the man, constituting the two sides of him that are going to meet in a symbolic, heterosexual union. Humboldt's vision of a gender-transgressing masculinity is more interesting than Goethe's, even though it is as problematic and based on a similar conception of sexual difference. In comparison with Goethe, Humboldt's vision of androgyny and *Bildung* is in fact more firmly rooted in the two-sex model.

For Humboldt, *Bildung* is not about resisting sexual ambiguity, but rather creating within the male soul a mating of female and male. Where *Wilhelm Meister* was retained its ambiguity about the two-sex model, it was a premise in Humboldt's two texts on sex and gender from 1794 and 1797. That they are based on a two-sex model is evident from their titles: 'Über den Geschlechtsunterschied' ('On sexual difference') and 'Über die männliche und weibliche Form' ('On the male and female form'). They were both published in Schiller's magazine *Die Horen*. They were not particularly well received, and Humboldt never published there again. The Humboldt scholar Paul Sweet writes that 'Humboldt...decided to begin with an examination of what the differences between the sexes signify in terms of a theory of the aesthetic *Bildung* of man'.[54] Sweet suggests that during this period Humboldt was particularly interested in 'the characteristics of the female sex'.[55] The shift to the two-sex model in combination with the 'outpouring' of books on the female sex during this period, as Petschauer mentioned, makes Humboldt's interest very much at one with the times. Humboldt was more open to women and questions of sexuality than Schiller, and did not believe that a husband should dominate his wife. Even though Schiller more clearly characterized the difference between the sexes, Humboldt shared many of the same views on women.[56] Sweet writes

54 Paul R. Sweet, *Wilhelm von Humboldt: A Biography* (Columbus: Ohio University State Press, 1978), 163.

55 Sweet 1978, 170.

56 Petschauer 1989, 280.

of 'Caroline Schlegel, who wrote that she and her friends had almost fallen out of their chairs laughing at Schiller's *Hausfrau* types, sarcastically called such creatures "versified Humboldtian femininities"'.[57]

In his two texts on sexual difference, Humboldt preached biological necessity. In the first essay, 'On sexual difference', he positioned sexual difference as the cornerstone of the natural order, making the two-sex model the grounds for everything. Sexual difference is the energizing force that alone defeats inertia:

> Without it [sexual difference] nature would not be nature, its machinery would stand still, and both the tension that unites all beings and the struggle that necessitates each individual arming himself with his own form of energy would cease, if a boring and enervating uniformity took the place of this difference.[58]

As MacLeod points out, Humboldt here sets about exploring and defining the physical world in sensualistic, aesthetic terms. Humboldt confirms sexual difference as a biological necessity at the same time as he views the androgynous union of polar oppositions as the highest state to be attained by humanity.[59] To McLeod's mind, Humboldt's use of words such as generation and conception (*zeugen* and *empfangen*) contains a 'utopian call for transcendence through artistic androgyny'.[60] I cannot agree with McLeod that Humboldt's references to heterosexual metaphors such as conception hold out the promise of transcendence. Compared to Goethe's gender-transgressing castrati, Humboldt's transcendence never goes outside the two-sex model and its implications of strict heteronormativity.

Using the metaphors of sexual intercourse, Humboldt imagined artistic creativity and *Bildung* as an erotic union. 'Not sufficient in itself, despite its riches, energy senses something outside itself—only when united with that other can it form a perfect whole....Once the imagination of the artist has given birth to a living image, the masterpiece is complete'.[61] For Humboldt, it is the artist, not the artwork, who can embody this androgynous ideal. Using sexually charged metaphors, he pictures how someone's male, rational side might fertilize their female, more sensitive side. For Humboldt, femininity existed *in* the man.

57 Caroline Schlegel to Louise Gotter, 10 February 1796, printed in Sweet 1978, 170.

58 Humboldt 1960, 268, quoted in translation in MacLeod 1998, 47.

59 MacLeod 1998, 47.

60 MacLeod 1998, 47.

61 Humboldt 1960, 276, quoted in translation in MacLeod 1998, 48.

Masculinity and femininity were two sides to man that were going to meet in a symbolic heterosexual union.

Humboldt moves on from the nature of the artist to explaining sexual difference itself. Like Rousseau, Humboldt perceives femininity as passivity and receptivity.

> This then is where sexual difference begins. The generative power is more suited to activate, the receptive power to be activated. What is given life by the former we call *masculine*, what imbues the latter with spirituality we call *feminine*. All that is masculine displays more independence, all that is feminine more suffering receptivity.[62]

Masculinity is coded as active and femininity as passive, and the feminine has greater 'suffering receptivity'—it is more of a victim of its gendered status. Women's more sexual character cannot accomplish the correct balance of masculine and feminine. These lines show how deeply connected the two-sex model of sexual difference was to hierarchal power structures.

Humboldt's metaphors of sexual intercourse also illuminate that sexual difference is activated at the very heart of his *Bildung* project. Petschauer, who is generally critical of Humboldt's views on women's education, interprets these lines as insisting on 'an ideal that both sexes should strive to attain'.[63] But since all that is masculine displays greater independence according to Humboldt, it is difficult to see how this could be read as an equal opportunity open to both sexes. Koselleck too claims that '*Bildung* was the legitimate place were Jews and women shared in equal rights and, even more so, was where they could take initiative. The Berlin salon culture produced *Bildung* at the same time as it generated the new emancipatory concept of *Bildung*'.[64] Yet Koselleck does not explain how this new emancipatory *Bildung* was open to women (or Jews), other than at a salon such as Rahel Varnhagen's. It was not until 1870 that German women were allowed to attend university lectures, and they were not permitted to matriculate and take university exams until 1900.[65] And as both Petschauer and Albisetti show, Humboldt had no interest in women's education; rather, the 'implications of his views for women's activities in the public

62 Humboldt 1960, 277, quoted in translation in MacLeod 1998, 48.

63 Petschauer 1989, 281.

64 Koselleck 2002, 182.

65 Michele Tournier, 'Women and Access to University in France and Germany 1861–1967', *Comparative Education*, 9/13 (Oxford: Carfax 1973), 107.

sphere were decidedly negative', Albisetti says.[66] He goes on, quoting Humboldt: 'Induced by their nature itself more to turn inward than to wander in the wide world, all receptive beings are chained to a steadier, less changeable course'.[67]

Koselleck refers to the new concept of love and marriage that was developed by women, but fails to notice, however, that the critique of marriage was a central element in the feminist legacy from writers such as Mary Wollstonecraft, Mme de Staël, George Sand, and many others before and after them. The reformulation of marriage on grounds of love is intrinsically linked to a notion of equality in reason, spiritually as well emotionally.[68] As we have seen, this was not the case of the marriage plot in *Wilhelm Meister* (or in any other novel by Goethe). It seems as if the kind of emancipatory *Bildung* for women that Koselleck refers to is just another word for feminism, and quite contrary to the *Bildung* envisaged by Humboldt and Goethe. Perfected *Bildung* was for them exclusively reserved for men. It is clear that Humboldt's view was that women should not transcend their femininity and take on male attributes or male occupations:

> The female sex on the other hand, must take the utmost care to preserve precisely this feminine identity, so that it does not destroy the vital expression of its own form; and if this effort completely fails, it will sink into its natural functions and into the routines of everyday life, or it will take on occupations that are really not part of its sphere.[69]

Masculinity in woman is a problem; and most of all, femininity in woman is a weakness. Or as Humboldt bluntly put it, 'The greater independence from sexual difference is a vital component of male *Bildung*'.[70] Ultimately, as McLeod writes 'it is woman's lot to remain constrained by biology, while men go on to attain higher *Bildung*, and "Menschlichkeit"'.[71] Humboldt's belief in the hierarchical relation between the sexes mirrors Rousseau's view of women as more dependent on their bodies. Petschauer writes that Humboldt, together with

66 Albisetti 1988, 5.

67 Albisetti 1988, 15.

68 See Claudia Lindén, *Om kärlek: Litteratur, sexualitet och politik hos Ellen Key* (Stockholm/Stehag: Brutus Östling bokförlag Symposion, 2002) ch. 1 & 2; and, for example, Niklas Luhmann, *Love as Passion: The Codifying of an Intimacy* (Stanford: Stanford University Press, 1998).

69 Humboldt 1960, 335, quoted in translation in MacLeod 1998, 51.

70 Humboldt 1960, 306, quoted in translation in MacLeod 1998, 51.

71 MacLeod 1998, 48.

Kant, Goethe, and Schiller, 'fully elaborated' on Rousseau's ideas.[72] If we think of how this argument was used against women during the French Revolution, it is obvious that it had reactionary implications for gender politics.

In terms of gender equality and women's education, Humboldt was a typical conservative of his time. Yet, as with Goethe, it is fascinating to read his reflections on sexual differences as an attempt to respond to a cultural situation that underwent a sea change in its gender discourse. In Humboldt's view, sexual difference stemmed from the then recently established two-sex model. But in this situation he sought to develop a hermaphroditic discourse of sorts, according to which it was still possible for a man to somehow harbour both sexes within himself. Humboldt's spiritual ideal ultimately becomes a union of masculinity and femininity, of perfect balance between opposites. In his essay 'On the male and the female form', Humboldt writes:

> The highest and perfect form of beauty demands not merely union, *but the most exact balance* of form and matter, of art and freedom, of spiritual and sensual unity, and this is only achieved when characteristics of both sexes are melted together in thought, and humanity is fashioned from the most intimate union of pure masculinity and pure femininity.[73]

As McLeod points out, what becomes clear in this quote is that androgyny is just a *balance*, not a fusion. Mixing the opposites does not lead to a new, third type of state; androgyny still preserves gender difference. Since McLeod's focus is androgyny, she does not ask what this kind of inner balance in man means for the production of masculinity. Humboldt does not transgress gender, but proper masculinity still contains femininity.

In the same way as the castrato's ability to perform femininity made him an artist, this capacity of balancing male and female became the epitome of creativity, of *Bildung*, in Humboldt's texts. It is connected to *Bildung* through the procreative metaphor, where masculinity always fertilizes femininity. In the late eighteenth century, men such as Goethe and Humboldt could still refer to their feminine sensibilities and in that way be perceived as *more* male. By the latter half of nineteenth century, femininity in a man had taken on connotations of homosexuality, and masculinity was seen in contrast to what was

72 Petschauer 1989, 272.

73 Humboldt, 'Über die männliche und weibliche Form', in Humboldt 1960, 296 quoted in translation in MacLeod 1998, 48. My emphasis.

un-male, thus making a man's femininity a threat to his maleness.[74] In queer theory, 'female masculinity' refers to women who take on masculine attributes.[75] With Humboldt, we should perhaps talk of 'male femininity', a certain type of femininity only available to men, in the same way as seen in the Neoclassical art in France. Humboldt's idea of masculinity in respect of *Bildung* is reminiscent of the ephebe's mixing of male and female characteristics.

To Solomon Godeau, the ephebe is a fetishistic figure that simultaneously acknowledges and denies sexual difference. She stresses that the currency of both the feminized masculine and the masculinized masculine in visual culture are jointly shaped by the expulsion of women from the public and cultural spheres.[76] 'Even as femininity was progressively defined in terms of a strict and biologically grounded difference, cultural production testified continually to *an internal division, a pervasive and haunting difference within the same*'.[77] Solomon-Godeau's examples of the popular ephebe show that this interest in the feminized masculine had a broad base in the culture that was linked to the expulsion of women from the public sphere.

When women fought for their rights to be citizens on equal terms with men, one aesthetic answer was to portray men and masculinity as androgynous, with a difference within. Who needs women if men already have femininity? As with Goethe's vision of an all-male theatre, the mixing of masculine and feminine in Humboldt's vision of rounded humanity had positive connotations for men but not for any actual historical women. Humboldt's vision of a masculinity that contained its own complementary femininity thus ultimately implied that the full potential of *Bildung* can only be realized by a transgendered man, even to the extent that this could be thought the ideal form in which humans can reach fulfilment. In other words, it takes a real man to show true femininity.

---

74 That masculinity is constructed in relation, not to women, but to what is un-male, is an important argument in masculinity studies. Classic texts in masculinity studies are R.W. Connell, *Masculinities* (Cambridge: Polity Press, 2005); G.L. Mosse, *The Image of Man: The Creation of Modern Masculinity* (New York: Oxford University Press, 1996).

75 Judith Halberstam, *Female Masculinity* (Durham: Duke University Press, 1998).

76 Solomon-Godeau 1997, 139.

77 Solomon-Godeau 1997, 141, my emphasis.

PART 2

Transformations of a Tradition

CHAPTER 4

Humboldt the Undead

Multiple Uses of 'Humboldt' and His 'Death' in the 'Bologna' Era[*]

Mitchell G. Ash

In public debate on higher-education reform in Germany today, the name of Wilhelm von Humboldt still serves as a marker for the classical German university ideal. The symbolic power of the name of Humboldt is acknowledged by both traditionalists and critics, and also by many specialists in higher-education history and policy research.[1] Historical research has challenged many of the claims about the German university tradition embodied by the name of Humboldt for some time, but this has had remarkably little impact on public discourse until very recently. In this chapter, however, I will not engage yet again in detailed refutation of these commonplaces of public discourse. Instead, I propose to reorient the discussion.

The focus here will be on recent representations of what has come to be called the Humboldt myth,[2] or 'Humboldt' (in scare quotes) in public

* Revised and expanded version of part of the plenary address 'From Berlin to Bologna and beyond: Humboldt and "Humboldt", past and present' presented at the conference 'The Humboldt Tradition', University of Uppsala, 25 November 2010. Unless otherwise noted, all translations from the German are my own.

1 Of course this is true outside Germany as well, see Thorsten Nybom, 'The Humboldt Legacy: Reflections on the Past, Present and Future of the European University', *Higher Education Policy*, 16 (2003), 141–159.

2 Walter Rüegg, 'Der Mythos der Humboldtschen Universität', in Mathias Krieg & Martin Rose (eds.), *Universitas in theologica. Theologica in universitatae: Festschrift für Hans Henrich Schmit zum 60. Geburtstag* (Zurich: Theologischer Verlag, 1997), 155–174; Mitchell G. Ash (ed.), *German Universities: Past and Future, Crisis or Renewal?* (Oxford: Berghahn Books, 1997) translated as *Mythos Humboldt: Vergangenheit und Zukunft der deutschen Universitäten* (Vienna: Böhlau-Verlag, 1999); id., 'Bachelor of What, Master of Whom? The Humboldt Myth and Transformations of Higher Education in Germany and the US', *European Journal of Education*, 41 (2006), 245–267; Sylvia Paletschek, 'Verbreitete sich ein "Humboldtsches Modell" an den deutschen Universitäten im 19. Jahrhundert?', in Rainer C. Schwinges (ed.), *Humboldt International: Der Export des deutschen Universitätsmodells im 19. und 20. Jahrhundert* (Basel: Schwabe, 2001), 75–104; id., 'Die Erfindung der Humboldtschen Universität: Die Konstruktion der Humboldtschen Universitätsidee in der ersten Hälfte des 20. Jahrhunderts', *Historische Anthropologie*, 10 (2002), 183–205.

 | DOI 10.1163/9789004271944_006

discussion of higher education in Germany. Before proceeding, some remarks on the meaning of the term 'myth' are in order. In popular, journalistic, and also scholarly discourse, myth is opposed to reality or truth. Myths are inventions meant to justify the positions of political power or social prestige occupied by their propagators, and often insisted upon despite evidence to the contrary precisely for that reason.[3] Though such an interpretation has its uses, it seems to me to miss a central point: 'Humboldt' embodies a cultural code, an invented tradition that has shown astonishing vitality despite, perhaps even because of, its multiple disconnections with actual university life, due to its multiple functions within academic culture and in the self-representation of universities in the public sphere.

A myth in this sense is not a lie, and its perpetuation in the face of contrary evidence need not only be due to stubborn adherence to an outdated ideology. Such an interpretation does not explain why a given myth appears to frame discourse so fundamentally that even critics of the power relations it is supposed to sustain appear unable to escape it. As I describe it here, the Humboldt myth is a culturally formative origin story, rather like the tales told by the peoples studied in Bronislaw Malinowski's classic, *Argonauts of the Western Pacific*, about how they came to live on their islands, or the story of Prometheus and the origins of fire.[4] The role of such tales is not to tell some sort of literal, historically verifiable 'truth', but to create, establish, and validate cultural meaning—that is, to establish the identity and self-worth of a given culture; in our case, an academic sub-culture—by means of historical narrative.[5]

3 Paletschek, in 'Die Erfindung der Humboldtschen Universität', appears to work largely within the common meaning of the term 'myth'. From this standpoint she criticizes what she takes to be my argument and that of Rüdiger vom Bruch ('A slow farewell to Humboldt? Stages in the history of German universities, 1810–1945', in Ash 1997, 3–27), for not deconstructing the 'Humboldt myth' radically enough, and therefore claims pride of place for her own, supposedly more radical deconstruction. Readers of my work and hers will realize that I disagree with Paletschek's views, but perhaps not in the way that she writes; unfortunately, space does not permit more detailed discussion here.

4 Bronislaw Malinowski, *Argonauts of the Western Pacific: An Account of Native Enterprise and Adventure in the Archipelagoes of Melanesian New Guinea* (London: Routledge, 1922), repr. in id., *Collected Works*, ii (London: Routledge, 2002). See also Percy S. Cohen, 'Theories of Myth (Malinowski Lecture 1969)', *Man*, 4/3 (1969), 337–353. For a profound analysis of the Prometheus tale and the cultural meanings of myth, see Hans K. Blumenberg, *Arbeit am Mythos* (Frankfurt am Main: Suhrkamp, 1979).

5 For a recent analysis of the cultural functions of such origin stories, see Albrecht Koschorke, 'Zur Logik kultureller Gründungserzählungen', *Zeitschrift für Ideengeschichte*, 1/2 (2007), 5–12.

Seen in this light, myths are located beyond the reach of truth and falsehood, in a place from which a given 'we'-grouping claims to have come, as part of a story that in the best case also explains where that group plans, ideally, to go or to remain. This meta-positioning of myth as a foundation story explains why groups with opposing power interests can and do participate in the tale, allowing it to structure their own self-consciousness as members of a given culture (or sub-culture), even as some of them claim to dissent from its nominally foundational principles. The historical meanings of such culturally formative stories can, indeed must, be studied for their own sake, whatever relation they may or may not have to the accounts historians construct of what went on in the past.[6]

The history of the Humboldt myth from its emergence around 1900 until the 1960s has been reconstructed in detail elsewhere.[7] In this chapter I consider more closely what I would like to call, not entirely jokingly, 'Humboldt the undead' in the context of another discursive formation—'Bologna'. By this I mean the many roles and functions of 'Humboldt', and particularly the often alleged 'death' of 'Humboldt' in recent debates on the European higher-education reform programme that began in the late 1990s and continue today. My thesis here is that the use of 'Humboldt' and his alleged 'death(s)' in current German debates mark the presence of an absence: the continued lack of a cultural code powerful enough to replace the one organized around 'Humboldt'. This lack I trace to a significant deficit in the Bologna Process, namely the failure of its architects and the continuing failure of its executors to formulate, or even consider, a vision of the meaning of higher education in the twenty-first century that goes beyond the logic of economic or administrative reason. I conclude with a brief consideration of whether 'Humboldt' the invented tradition still has, or should have, any meaning in the twenty-first century.

Meanings of 'Humboldt' in Germany before and after Reunification

The rapid expansion of German higher education since the 1960s has had a profound impact on the 'Humboldt' discourse. For old-school conservatives, the mass university as such meant, and still means, the 'death of Humboldt'.

---

6 For an interpretation of Livius' classic history of Rome along similar lines as a 'retrofiction', see Albrecht Koschorke, Susanne Lüdemann, Thomas Frank & EtelMattala de Mazza, *Der fiktive Staat: Konstruktionen des politischen Körpers in der Geschichte Europas* (Frankfurt am Main: Fischer Taschenbuch, 2007), pt. 1 esp. 51f.

7 See *inter alia* the literature cited above, n. 2.

There is a certain justification for this claim, given that universities in Humboldt's time were indeed elite institutions that only educated a tiny percentage of a given age cohort; however, this traditionalist line of 'Humboldt' defence and attack on mass higher education was not shared by modernizing conservatives or liberals. Modernizers such as Georg Picht and Ralf Dahrendorf accepted the expanded university as a source of qualified labour for science-based industries and the civil service of an expanding welfare state.[8] On the Left, the predominant position could be described as 'Humboldt' for everyone. For the education reformer Ludwig von Friedeburg, for example, the idea on which Wilhelm von Humboldt based his proposal for a general reform of the Prussian education system—that all men should be educated for enlightened citizenship, irrespective of their later occupations in life—made him an early proponent of unified secondary education (the *Gesamtschule*).[9] The central point here, however, is that higher-education policy debates in West Germany after the 1960s did not only take place between left-wing advocates of maximum access and conservative elitists, but also within the West German conservative and centrist–liberal camps, between old-school elites and 'modernizers'.

In practice, after the so-called 'opening resolution' of 1977 allowed practically unlimited university enrolment by any student with the correct qualifications (on the self-serving, wildly wrong assumption that this would produce only temporary overcrowding), the structural underfunding of higher education became accepted policy throughout the Federal Republic. Studies from the period left little doubt that, despite the large number of newly founded institutions, the expansion of higher education was being financed primarily by overburdening the existing infrastructure. Between 1970 and 1987, expenditure per student, corrected for inflation, dropped by one-third from DM21,600 to DM14,800.[10] Because student numbers expanded far more rapidly than those for faculty, faculty–student ratios worsened by 50 per cent, from slightly less than 1:40 in 1972 to more than 1:59 in 1990.[11] If 'Humboldt' continued to

8 Konrad Jarausch, 'The Humboldt Syndrome: West German universities 1945–1989—an Academic *Sonderweg*?', in Ash 1997, 33–49 at 38f.

9 Ludwig von Friedeburg, *Bildungsreform in Deutschland: Geschichte und gesellschaftlicher Widerspruch* (Frankfurt am Main: Suhrkamp, 1992), 64f.

10 Paul Windolf, *Die Expansion der Universitäten 1870–1985: Ein internationaler Vergleich* (Stuttgart: Enke, 1990), 237.

11 Figures from the Statistisches Bundesamt, summarized in Wissenschaftsrat, *Empfehlungen zur Qualitätsverbesserung von Lehre und Studium* (Cologne: Wissenschaftsrat, 2008), 23, fig. 1.

pace the halls of alienating concrete university buildings in Bochum, Bielefeld, Frankfurt, Cologne and elsewhere, it was as a living corpse, a reminder of what some at least continued to imagine or wish universities to be, while seminars in such subjects as history or German literature grew to outlandish size.[12]

In reaction to this fraught situation, coping strategies and organized irresponsibility became the norm. The allegedly Humboldtian idea of open admission for all students with an appropriate school leaving certificate (*Abitur*) from a gymnasium became a legal claim to a place limited only by institutional capacity; overcrowded programmes in fields such as medicine and psychology were allowed to introduce enrolment limits (*numerus clausus*). The status of professors as civil servants supported the convenient fiction that all universities should not only be accorded equal standing in law, but should also be considered equal in fact, regardless of differences in their actual quality. 'Freedom of learning' for many students meant maximizing their time to degree completion—or high dropout rates—while 'freedom of teaching' for many university teachers meant studied indifference to the learning needs of all below the level of postgraduate research student, or to the demands of the labour market.[13] 'Humboldt'—meaning in this case the ideal of research-oriented teaching—remained present, if at all, in upper-level seminars, in smaller disciplines (the continued existence of which is permanently under threat), and in postgraduate studies. American observers took note. Daniel Fallon described Humboldt's ideal in 1980 as being 'a heroic ideal in conflict with the modern world'.[14] Nonetheless, 'the unity of teaching and research' continued to be upheld as an identity-building slogan, despite its utter lack of meaning in practice in the more heavily subscribed subjects.

Ironically, it was in the 1970s, and not earlier, that 'Humboldt' as a symbol for the German university ideal became a standard trope in the speeches of German university rectors and presidents.[15] Why did the Humboldt myth retain its appeal for decades after the rise of the mass university? First, of

12 For the view that 'Humboldt' lived on in the humanities, see Dieter Langewiesche, 'Humboldt ist lebendig: Geisteswissenschaften in der Massenuniversität', *Frankfurter Allgemeine Zeitung*, 21 December 1995.

13 For summaries of these widely known crisis diagnoses, see Jarausch, 'The Humboldt syndrome', in Ash 1997, and Peter Lundgreen, 'Mythos Humboldt Today: Teaching, Research, Administration', in Ash 1997, 127–148.

14 Daniel Fallon, *The German University: A Heroic Ideal in Conflict with the Modern World* (Boulder, CO: Westview Press, 1980).

15 Dieter Langewiesche, 'Humboldt als Leitbild? Die deutsche Universität in den Berliner Rektorats seit dem 19. Jahrhundert', *Jahrbuch für Universitätsgeschichte*, 14 (2011), 15–37.

course, it was and remains a symbol for the corporate autonomy and status-consciousness of full professors; interestingly, this appears to be the case also for professors who think of themselves as politically progressive. But this is not the whole story. Second, 'Humboldt' is a symbol for research and teaching oriented toward aims that go beyond immediate financial gain or short-term social usefulness. This ideal, so attractive in the 1970s, remains so today for both old-school conservatives and progressives who continue to advocate higher education's potential as a free space for socially critical thought and open debate. And here myths can help establish what is now called 'corporate identity', even if—or perhaps even because—the ideals in question are utopian.

A fundamental shift in this constellation began after 1990. At first, German reunification appeared to strengthen the power of 'Humboldt', in this case as a double-edged symbol. In the new federal states of eastern Germany, 'Humboldt' was and remains for many a symbol of freedom.[16] During the reunification process itself, however, 'Humboldt' became a sign for the restructuring of East German universities, in particular their staffing, according to West German models. This came at a very high price, as thousands of mid-level teaching staff who had tenure before 1990 were dismissed or put on short-term contracts and then pensioned off, because positions for them no longer existed.[17] Once the West Germanification of the East German higher education and research systems was more or less complete, the time was ripe to revive long-standing reform agendas of a different kind.

Detlef Müller-Böling, from 1994 to 2008 head of the Centre for Higher Education, a think-tank financed by the Bertelsmann Foundation in Gütersloh, outlined this turn in a book entitled *The Unchained University*.[18] Müller-Böling proposed a twofold differentiation: (*i*) a clear distinction among types of

---

16 This appears to have been the case throughout the former Communist bloc. See, for example, Guy Neave, 'On Scholars, Hippopotami and von Humboldt: Higher Education in Europe in Transition', *Higher Education Policy*, 16 (2003), 135–140 at 137f.

17 For the catastrophic results see Mitchell G. Ash, 'Higher Education and German Unification: "Renewal" or the Importation of Crisis?' in Ash 1997, 84–109; and Konrad H. Jarausch, 'Säuberung oder Erneuerung? Zur Transformation der Humboldt-Universität 1985–2000', in Michael Grüttner, Rüdiger Hachtmann, Konrad H. Jarausch, Jürgen John & Matthias Middell (eds.), *Gebrochene Wissenschaftskulturen: Universität und Politik im 20. Jahrhundert* (Göttingen: Vandenhoeck & Ruprecht, 2010), 327–352.

18 Detlef Müller-Böling, *Die entfesselte Universität* (Gütersloh: CEH, 2000); id., 'Entfesselung von Wettbewerb: Von der Universität zum differenzierten Hochschulsystem', in Grüttner et al. 2010, 353–377.

higher-education institution in Germany between full or technical universities and institutions oriented more narrowly towards occupational training (*Fachhochschulen*), in order to reduce the predominance of universities; and (*ii*) a distinction between high-end, research-oriented institutions and other, more teaching-oriented universities. The achievement of yet a third differentiation, between 'world-class' and other research universities, is the goal of the 'Excellence Initiative' of the Federal and *Länder* governments that began in 2007.[19] The GATT and the Treaty of Lisbon were clearly part of the wider context of this transformation; the alleged 'marketization' of higher-education policy and the effort to transform universities into business-style operations with the help of new public management has been analysed—and criticized—extensively.[20] But references to 'marketization' and to oft-repeated OECD recommendations to increase the number of academics in response to the demands of knowledge society cannot alone explain the widespread and remarkably rapid acceptance of this new policy idiom.

'Humboldt' and 'Bologna', or: Humboldt the Undead

'Humboldt' has died many deaths, and been resurrected equally often; it is tempting to tell this tale as a zombie movie. There have been many murderers, and equally many resurrectionists; sometimes they were the same people. In the past few years, this narrative trope has become linked with another trope: 'Bologna'.

In a speech to the German Rectors' Conference in 1997, Jürgen Rüttgers (CDU), then 'Future Minister' in the government of Chancellor Helmut Kohl, caused a stir by proclaiming 'Humboldt is dead!'. Rüttgers was not an old-school, but a 'modernizing' conservative. The following year, in 1998, he joined his counterparts from France and Italy in signing the Sorbonne Declaration, and was also the minister primarily responsible for the passage of a revised Federal Higher Education Framework Act. An experimental clause in that law gave the German *Länder* hitherto unknown flexibility in the approval of course programmes and curricula. A recent study argues convincingly that Rüttgers's

19 For a recent analysis, see Stephan Leibfried (ed.), *Die Exzellenz-Initiative: Zwischenbilanz und Perspektiven* (Berlin: Berlin-Brandenburgische Akademie der Wissenschaften, 2010).

20 Space does not permit an extended discussion of this policy transformation. For discourse analyses of the new policy-speak see, for example, Bob Jessop, Norman Fairclough & Ruth Wodak (eds.), *Education and the Knowledge-Based Economy in Europe* (Rotterdam: Sense, 2008).

signature on the Sorbonne Declaration was intended to provide international flanking support for his domestic reform agenda.[21]

The Bologna Declaration, with its well-known aim of creating a European higher-education area, within which all participating countries would approve modularized curricula based on two—later three—main cycles of undergraduate and postgraduate study, came in 1999, after these reforms had already been enacted. The political, economic, and technocratic forces driving the Bologna Process have been analysed extensively elsewhere.[22] Relevant here is the role of 'Humboldt'; more precisely, of his alleged 'death' at the hands of the 'Bologna' discourse.

One response of old-school conservatives to Rüttgers's proclamation of Humboldt's 'death' in 1997 came from Hartmut Schiedermair, President of the German Higher Education Association, an organization representing university teachers. In a 2001 address in Erfurt, he asked whether the Humboldtian ideal of the university was dead, and responded with a list of well-known complaints about mass higher education; the conversion of science and scholarship into 'purchasable goods'; the corresponding emphasis on third-party funding levels as a measure of academic quality; and bureaucrats' ignorance of Humboldt's original ideals, especially the freedom of professors to organize university affairs and appoint their own colleagues. To this he added a warning that planned changes in the calculation of professors' salaries would lead to 'organized mediocrity' and a 'farewell to the Humboldt university'.[23] Schiedermair was seconded by another Bavarian, Heike Schmoll, an editor of the *Frankfurter Allgemeine Zeitung*, who published numerous polemical editorials critical of 'Bologna' and then summarized her critique of the mass

21 Pauline Ravinet, 'The Sorbonne Meeting and Declaration: Actors, Shared Vision and Europeanisation', in Tor Halvorsen & Atle Nyhagen (eds.), *The Bologna Process and the Shaping of the Future Knowledge Societies, Conference Report from the Third Conference on Knowledge and Politics* (Bergen: University of Bergen, 2005), 187–204; and Cornelia Racké, 'The Emergence of the Bologna Process', in Halvorsen & Nyhagen 2005, 205–213.

22 See the chapter by Susan Wright in this volume. In my view, analyses from the Left that focus strongly on GATT, the Lisbon Treaty, neo-liberal 'knowledge society' discourses, and the like make important points, but pay too little attention to the autonomous role of technocratic discourse, with its emphasis on improving the so-called 'productivity' (or 'throughput') of the system, including the reduction of dropout rates—something that many on the Left might also wish for—and quality control. Still less attention has been paid to the role of student demand (see below).

23 Hartmut Schiedermair, 'Ist die Universitätsidee von Wilhelm von Humboldt tot?', in Wolfgang Bergsdorf (ed.), *Erfurter Universitätsreden 2001/2002* (Munich: Iudicum, 2002), 9–25 at 10 and 11.

university in a book, *Lob der Elite* ('In Praise of Elites'), in 2008.[24] Clearly, the struggle between old-school and modernizing conservatives has continued unabated since the 1960s.

In an interesting twist, the old-school conservative position soon received eloquent support from the Left. In a best-selling book, *Theorie der Unbildung* ('Theory of Non-Education'), the Viennese philosopher Konrad Paul Liessmann drew a direct line to 'Bologna' from Theodor Adorno's 1959 critique of the 'half-educated' citizenry produced by the mass media and the culture industry. In Liessman's view, 'Bologna' stands for the intentional dismantling of Humboldt's ideals and the end of the *Bildungsbürgertum* that had at least pretended to subscribe to them. Employing quotations from Humboldt's general writings on education as well as his memorandum on the university, Liessmann argued that 'the one-sided assessment of university careers according to research productivity and the degradation of teaching' to information transfer have 'sabotaged' the traditional unity of teaching and research. Under the guise of making education available to all and establishing certifiable skills that are transferable throughout Europe, the narrow-gauge, occupationally oriented 'Bologna' Bachelor degree curriculum 'demolishes the universities from within', transforming the meaning of *Bildung* from individual self-cultivation (achieved through science- or scholarship-based learning) to the arbitrary assemblage of uncoordinated contents and credit points, attesting to little beyond the time-management skills needed to acquire them.[25]

In 2009, mass student protests began at the Academy of Fine Arts in Vienna and then spread to Germany. These outbursts made it abundantly clear that the realization of the technocrats' dream of a uniformly structured European higher-education area was not proceeding smoothly. There were multiple divisions within the protest movement, the main one being between advocates of a work-load protest and a more radical group affiliated with the

24 Heike Schmoll, *Lob der Elite* (Munich: C.H. Beck, 2008). Schmoll's position is not reactionary; she argues, rather, that democracies need elites in order to function.

25 Konrad Paul Liessmann, *Theorie der Unbildung: Die Irrtümer der Wissensgesellschaft* (Munich: Piper, 2008), 118 & 119. Interestingly, Liessmann does not denounce this as Americanization, as others have done. Other contributions to the debate from the Left were structured largely as critiques of neo-liberal 'elite' or 'excellence' discourses and of the 'entrepreneurial university', with fewer direct references to 'Humboldt': see, e.g. Richard Münch, *Die akademische Elite: Zur sozialen Konstruktion akademischer Exzellenz* (Frankfurt am Main: Suhrkamp, 2007); and id., *Akademischer Kapitalismus: Zur politischen Ökonomie der Hochschulreform* (Frankfurt am Main: Suhrkamp, 2011).

anti-globalization movement that rejected 'Bologna' altogether as part of a neo-liberal plan to subjugate freedom-loving students to the dictates of the market. In these students' critiques, the 'death of Humboldt' motif took on new meanings. An example is the online article from August 2009 with the title 'Humboldt is dead', in which the unnamed student author alleged that central points of the Bologna Process were formulated directly by the European Round Table of Industrialists, and concluded from this—rather like the old-school conservative Hartmut Schiedermair—that 'with the Bologna Process, education is being turned into a consumer item'. Interestingly, however, the author followed this with a very explicitly formulated work-load protest: 'It is practically impossible to store in one's brain for any period of time the contents of all of the modules that one has to complete in order to advance, since ever less time remains. Hardly any content gets into students' long-term memory; instead short-term memory is being trained'. The article ends with a quotation from Humboldt suggesting that policymakers who think only in quantitative terms should be regarded with suspicion for wanting 'to make machines out of men'.[26]

Whether such views were common among the protestors, much less among students more generally, is doubtful. Indeed, the vast majority of students—and their parents—might well be the OECD's best allies in this struggle, since their primary concern is to complete their studies as quickly as possible and emerge with degrees that qualify them for better-paying jobs.[27] The problem in this particular case was that not the OECD or European industrialists had killed 'Humboldt', but that university faculties themselves had done the deed by accepting the three-year limit for the Bachelor's degree and then creating overstuffed and unnecessarily rigid curricula.[28] In the process, 'Bologna' became synonymous with school-like study programmes of questionable value to the labour market, even though no international policy adopted since 1999 actually mandated such curricula.

---

26 'Humboldt ist tot—Eine Kritik am Bologna-Prozess und dessen Auswirkungen', 26 August 2009, <http://veritascode.wordpress.com/2009/08/25/humboldt-ost-tot-%E2%80%93-eine-kritik>, accessed 14 November 2010. The quotation did not come from Humboldt's writings on education, but rather from his text on the limits of the state (with a wildly incorrect date).

27 On this point see the analysis in Uwe Schimank, 'Humboldt in Bologna—falscher Mann am falschen Ort?', in HIS (ed.), *Perspektive Studienqualität—Themen und Forschungsergebnisse der HIS-Fachtagung 'Studienqualität'* (Bielefeld: Bertelsmann, 2010), 44–61.

28 The internal power struggles within academic departments that produced such results deserve an analysis in their own right.

Goodbye 'Humboldt'?

In this context, the 'death of Humboldt' trope acquired yet another dimension: the historical 'Humboldt myth' now itself became the subject of more widespread reflection. In 2008 the *Deutsche Universitätszeitung* (DUZ), a widely read monthly magazine devoted to higher education, ran a series entitled 'Goodbye Humboldt? What values will the university need in 2010?' with an eye to the deadline set for the end of the first phase of the Bologna Process. In the first article in the series, entitled 'Die Einheit von Forschung und Lehre' ('The Unity of Teaching and Research'), the higher education researcher Peer Pasternack spelled out that 'the unity formula cannot explicitly be found in Humboldt's writings. It is a creation of later historiography'. He went on to add that academic study in the nineteenth century was neither free in the sense of being lacking in orientation towards economic or social goals, nor was it generally research-oriented.[29]

Soon after the anti-'Bologna' protests began, Jürgen Rüttgers weighed into the debate a second time. Speaking this time as Prime Minister of North Rhine–Westphalia, but again addressing the German Rectors' Conference, in April 2009 he modified his earlier 'death of Humboldt' proclamation in an interesting way: 'Research on loneliness and freedom, as Humboldt once imagined it, is no longer possible in a globalized knowledge society. Similarly, the unity of research and teaching reaches its limits in the age of the mass university, if it is merely proclaimed and not actually made possible. That unity has in recent years become a myth [...] It may seem paradoxical, but one must reinvent the Humboldtian university in order to save it [...] Humboldt's university is dead. Humboldt's ideal is and will remain alive!'[30]

With this text, references to the 'Humboldt myth' entered the public debate. In May 2009, Wilhelm Krull, long-time Secretary General of the Volkswagen Foundation, responded directly to Rüttgers in a way that made it clear that sophisticated policymakers had begun to co-opt the university historians' discussion. Krull devoted the first half of his paper to an extended, accurate summary of the work of Sylvia Paletschek on the history of the 'Humboldt myth', cited in the introduction to this chapter. Turning to the current situation, he acknowledged that 'catastrophic' faculty–student ratios in German universities (now 1:66, as against 1:9 at top British and American universities) 'place

29 Peer Pasternack, 'Die Einheit von Forschung und Lehre', *DUZ Magazin*, February 2008, 20f.

30 'Rede von Ministerpräsident Dr Jürgen Rüttgers anlässlich der Jahresversammlung der Hochschulrektorenkonferenz an der RWTH Aachen am Montag den 20 April 2009'.

narrow limits' on the realization of Humboldtian ideals. Surprisingly, he claimed that 'Humboldt' is experiencing 'an enjoyable Renaissance' precisely in the Bologna Process context, but hastened to add that these ideals should not be misused as a all-purpose weapon in defence of an elitist concept that was already outdated when it was first mythologized a century ago. Krull's suggestion was not to regard students as 'customers', but rather to take them seriously—at the postgraduate level and earlier—as co-producers first of their own learning and then of knowledge, and to revivify the unity of teaching and research in structured postdoctoral programmes: 'Not in an opposition, but in a collaboration of the Humboldtian ideal and Bolognese pragmatism lies the future of an increasingly differentiated science system'.[31]

The position of the Bielefeld sociologist of science Peter Weingart, located in the liberal middle of the debate, proved most insightful. Bringing the German Federal government's 'Excellence Initiative', mentioned above, into the discussion, he wrote that 'The current higher-education reform, associated superficially in the media with claims of "excellence" or with "Bologna", is considered by nostalgists to mean the end of the Humboldtian university model. But no concept has taken its place that could properly ground the function of higher-education institutions'; instead, the ideal of 'accountability' has been perverted to mean 'accounting' in the now 'entrepreneurial university'.[32] With this, Weingart put his finger on the real cause of the problem. Neo-liberal marketization and the discourse of administrative reason embodied in the jargon of new public management have failed to capture the deeper meaning or internal logic of higher education, or to inspire the relevant actors outside university central administrations.[33]

31 Wilhelm Krull, 'Hat das Humboldtsche Bildungsideal noch eine Zukunft? Impulsreferat zum Symposium 'Wissen und Geist. Universitätskulturen', Leipzig, 11–13 May 2009, esp. 9–11; see also id., 'Bildung und Wettbewerb', in Andreas Schlüter & Peter Strohschneider (eds.), *Bildung? Bildung! 26 Thesen zur Bildung als Herausforderung im 21. Jahrhundert* (Berlin: Berlin Verlag, 2009), 194–207. The student–faculty ratios cited by Krull are actually far worse in heavily oversubscribed subjects such as history, German literature, or information science (Wissenschaftsrat 2008, 22f and 26–33).

32 Peter Weingart, 'Humboldt im Ranking', in Schlüter & Strohschneider 2009, 249–259 at 250 & 257.

33 The old new Left took up the 'death of Humboldt' trope as well. Unfortunately, texts from this group added little of substance to the discussion. See, for example, Wolfgang Lieb, 'Humboldts Begräbnis. Zehn Jahre Bologna-Prozess', *Blätter für deutsche und internationale Politik*, June 2009, <http://www.blaetter.de/archiv/jahrgaenge/2009/juni/humboldts-begraebnis?print>, accessed 14 November 2010.

The celebration of the two-hundredth anniversary of the Humboldt-Universität zu Berlin in 2010 was an obvious occasion for reflection on the state of play. The large array of events presented in connection with the commemoration included, among others, a discussion series with titles like 'What can we still learn from Humboldt?' (*Was von Humboldt noch zu lernen ist*) or 'The future of universities in a world of research' (*Die Zukunft der Universitäten in einer Welt der Forschung*; the original title appeared first in English and then in German). The concluding event in the series was even entitled 'The reality of a myth' (*Die Wirklichkeit eines Mythos*). Here, too, the point that critical historians have been trying to make seems finally to have been understood: an unbroken 'Humboldt' story can no longer be told in the age of globalized mass education.

Most impressive among the many contributions was the collection of speeches by Christoph Markschies, an internationally renowned historian of the early Catholic Church and President of the Humboldt University in its jubilee year, published with the title *Was von Humboldt noch zu lernen ist* ('What is still to be learned from Humboldt').[34] Echoing Jürgen Rüttgers's statement, but with a new twist, he wrote: 'The Humboldt myth is dead, long live Humboldt!' To this he added the following theses: '*Universitas litterarum* does not mean many tiny specialisms in one building. *Universitas litterarum* 1810 [meant]: concern for the whole. Today [it means]: overcoming unproductive dualisms. Reform more carefully, and be guided by science and scholarship! Convert the leading German universities into non-profit foundations! End the rigid commitment to single-state universities! Finally, internationalize! In questions of teaching and research: more risk-taking!'[35]

Markshies's relatively critical, reformist stance will surely have surprised those of his colleagues who were looking for uncritical cheerleading of the sort usually offered by university presidents in jubilee years, or even for a call for a return to the grand ideals of bygone days. Yet the positions advanced here exemplify the tendency among contemporary German higher education and science managers towards the sophisticated co-opting of critical viewpoints rather than PR-orientated blindness towards them. Whether such sophistication can be translated into intelligent and effective reform policies remains to be seen.

---

34 Christoph Markshies, *Was von Humboldt noch zu lernen ist* (Berlin: BUP, 2010).

35 Ibid. 7 ff.; see also Harald G. Kratohvila, review of Christoph Markschies, 'Was von Humboldt noch zu lernen ist', <http://socialnet.de/rezensionen/9602.php>, accessed 23 November 2010; Frank Hahn, '200 Jahre Berliner Universität: Was ist von Humboldt noch zu lernen?', <http://solon-line.de>, accessed 23 November 2010.

Conclusion

At the beginning of this chapter I suggested, first, that myths need not only be lies told by ruling elites in order to safeguard their power—although they can surely be that—but are often shared narratives that create, establish, and validate identity and self-worth for many participants in a culture or sub-culture. I suggested, further, that the mythical tale embodied by the name of Humboldt has a history of its own, which began around 1900 and should be analysed not as an ideology that obscures the 'real' history of German-speaking universities (as well as the history of universities in countries that ostensibly imported the German model), but rather as a formative component of that history. In this chapter, I have attempted to sketch the most recent portion of that myth-history. Part of the story is the continuing power of 'Humboldt', despite, or perhaps even because of, the increasingly obvious discrepancy between that vision and the reality of mass higher education. Perhaps more fascinating, however, is the way in which at least some high-level higher-education and science policymakers in reunified Germany have recently tried to distance themselves at least partly from 'Humboldt' discourse. Indeed, some of these policy-makers have even picked up on recent historical scholarship pointing to the discrepancy between 'Humboldt' and 'real' university history, as well as to the historical character of the Humboldt myth itself. It is possible that this more sophisticated tale might even become the official story. Yet even Jürgen Rüttgers, perhaps the first to declare 'Humboldt' dead and to welcome the fact, could not bring himself to drop the famous name altogether. 'Humboldt' the zombie still walks among us.

Wilhelm von Humboldt the man is long dead, and the mythic tradition created in his name now appears to be in crisis in his own homeland. There are two reasons for this. First, the primacy of the claims of society on higher education is finally being acknowledged and accepted in German-speaking Europe, as it always has been in the US. As Peter Strohschneider, a professor of medieval German literature in Munich, former head of the German Science Council, and now President of the *Deutsche Forschungsgemeinschaft* (German Research Council), puts it, the German universities did not become as large as they did because of the expansion of scientific knowledge alone, but primarily due to the expansion of society's demand for higher education and the employment opportunities it is thought to provide.[36] There is little room left for the

36 Peter Strohschneider, 'Zu einigen aktuellen Entwicklungslinien des deutschen Wissenschaftssystems', in Grüttner et al. 2010, 367–377 at 372.

pursuit of *Bildung* in 'loneliness and freedom' in such a scenario. For Bremen sociologist Uwe Schimank, 'Bologna' thus represents the victory of middle-class or even petit-bourgeois careerism over elitist *Bildungsbürger* ideals.[37]

The second and much more important reason for this crisis of meaning is that technocratic slogans from the handbooks of new public management have failed to provide a suitably inspiring alternative to the utopian vision of 'Humboldt'. What I have called 'Humboldt the undead' marks the presence of an absence. Thus the historian Dieter Langewiesche writes that at a time when cultural traditions and scholarly canons have lost their force, 'A society with no idea of *Bildung* requires a university with no claim to provide it'.[38] Yet he goes on to suggest that Humboldtian '*forschendes Lernen*' (learning through research) can still be a significant qualification, precisely because the skills involved are not limited to any single subject or occupation.[39] This seemingly hopeful suggestion is in fact a capitulation: what was once supposed to be the royal road to truly meaningful higher education and self-directed personality formation would now seem to be just one fast track among many to certifiable skills.

To my mind, the invented tradition of 'Humboldt' in any case always was, and remains, a utopia, albeit an appealing one. Utopias can be useful as visions of what might be possible in ideal circumstances, or as goals to strive for. 'Humboldt for all', however, is not a utopia, but an illusion, and one not shared by the majority of students today or their parents. This fundamental point continues to be obscured by the pervasive, largely status-driven, and utterly false self-image of many university teachers in German-speaking Europe, especially though not only in the humanities and the social sciences. For many of these actors, 'Humboldt' still appears to have real meaning—as a nostalgic vision of their own education and an inadequate formulation of their own hopes for their role in the present. University teachers in Europe will need to develop a different and more realistic idea of their profession if higher education is to have any meaning beyond the mass production of certified 'academic' professionals.

As far as 'Bologna' is concerned, the creation of a European higher-education area remains the most important higher-education reform project of the past century, even if the result will not follow an American or even an 'Anglo-Saxon' model. The key issue remains the sustainability of the European welfare

---

37 Schimank, 2010.

38 Dieter Langewiesche, 'Chancen und Perspektiven: Bildung und Ausbildung', in id., *Zeitwende: Geschichtsdenken heute* (Göttingen: Vandenhoeck & Ruprecht, 2008), 219.

39 Ibid.; see also Pasternack 2008 (as above, n. 28).

state, and higher-education policy debates are a subset of that larger question. Where higher education continues to be regarded as the responsibility of the state, as is the case in Continental Europe and Scandinavia, tuition-free higher-education institutions (with open admissions, at least in Austria) are part of welfare state service provision, and thus compete for funding on highly unfavourable terms with other, far more expensive programmes such as pensions, health care, or housing subsidies. For Americans this is largely a non-issue; student subsidies and loans or state support research aside, higher education has never been regarded exclusively as a public good in that country.

In the optimistic scenario offered by the sociologist Rudolf Stichweh when he was rector of the University of Lucerne in 2008, the Bologna Process means acceptance, after more than forty years of resistance and denial, of the mass university and its imperatives as facts to be faced and dealt with pragmatically by students as well as by teachers and administrators.[40] Whether and how that will actually happen remains unclear. The choices have been on the table for some time: either adequate state funding plus sufficient financial and legal autonomy to enable the acquisition of additional funding without losing the state's contribution, or permanent scarcity leading to speed-up, burn-out, and deep frustration for both faculty and students. Regardless of the policy choices to be made, as far as the self-conception of the participants is concerned no alternative to the inspiring 'Humboldt myth' is presently in sight.

What the acceptance of mass higher education in Europe may mean for 'Humboldt' remains to be seen. University presidents continue to evoke 'Humboldt' in their public speeches for lack of a better alternative. And yet the realization appears to be dawning that the laissez-faire approach to teaching and learning in the mass university that in the past had been allowed to dominate under the banner of 'Humboldt' must give way to more carefully conceived curricula, more focused attention on teaching ability during university teacher-training and in the workplace, and increased efforts to establish a genuine contractual relationship between teachers and students in order to ensure that students actually have a chance to complete their degrees while still producing results of adequate quality. That is what the Bologna process is actually intended to achieve, after all. 'Bologna' now has a bad name, for reasons that have more to do with its highly unfortunate implementation than with its actual goals. Whether the new reforms that will surely be needed can or should still bear the 'Humboldt' label or not may not matter much in the end.

---

40 Rudolf Stichweh, 'Universität nach Bologna. Zur sozialen Form der Massenuniversität: Akademische Rede am Dies Academicus der Universität Luzern am 29. Oktober 2008', *Luzerner Universitätsreden*, No. 19.

CHAPTER 5

'Humboldt' in Belgium

The Rhetoric of the German University Model

Pieter Dhondt

Introduction

During the nineteenth and early twentieth centuries, the German university was hailed in Belgium, as elsewhere, as a model institution, existing solely to serve pure scholarship, with professors and students who gave and attended courses in complete freedom, without barriers between the faculties, or indeed between individual universities. Both the individual student and the university as a whole could develop in *Einsamkeit und Freiheit*, or completely independent from society. Its French counterpart, on the other hand, was administered entirely by an authoritarian central government, it was said. The German idea of *Einheit der Wissenschaften* would have been beyond the grasp of the French *écoles spéciales*. Scholarship was not central there; utilitarian vocational training was.

Of course, these stereotypes of university education in both countries were not totally unfounded, but they did not take into account geographical differences or shifts over time, nor the actual implementation of some specific characteristics of the model in daily practice. Nevertheless, in the nineteenth century, professors, politicians, and journalists in Belgium, much as in other countries, used *le système universitaire allemand/français* as a reference point, which, in time, achieved the same meaning for all of them.[1] 'It became commonplace in the nineteenth century to contrast French specialization and centralization with the relatively autonomous German universities, which combined teaching and research', as the Scottish historian Robert Anderson confirms.[2]

Although the German model of university education lost much of its attraction immediately after the First World War, particularly in war-ravaged Belgium, it nevertheless still continued to exert a certain influence worldwide, and the figure of Wilhelm von Humboldt as the founder of the University

1 Ulrich Teichler, *Europäische Hochschulsysteme: Die Beharrlichkeit vielfältiger Modelle* (Frankfurt am Main: Campus, 1990).

2 Robert D. Anderson, 'Before and after Humboldt: European Universities Between the Eighteenth and the Nineteenth Century', *History of Higher Education Annual*, 20 (2000), 5.

 DOI 10.1163/9789004271944_007

of Berlin gradually took on almost mythic proportions. The aim of this chapter is to show how Humboldt in particular and the German university model in general have functioned as rhetoric instruments whenever university education was discussed in Belgium from the mid nineteenth century onwards. The focus will be on the search for the origins of German supremacy and late nineteenth-century attempts to deal with it; the direct impact of German educational administration in the First World War through the establishment of the Dutch-speaking Von Bissing University in Ghent; and finally the increasingly common use of the name of Humboldt as a merely rhetorical instrument in recent years.

It will become clear that throughout the period in question, Belgian professors and administrators, like their peers elsewhere, always adapted the famous German university model to their indigenous system, and referred to it without taking sufficient account of its context in time and place. One of the largest specific challenges in the Belgian case was, and to a certain extent still is, how to reconcile the intrinsic unlimited freedom of the German university (at least according to the rhetoric) with the consequences of the extrinsic unlimited freedom of education in Belgium following the extremely liberal Constitution of 1831. The only way for the government to ensure that people had the right to establish universities in complete freedom, while continuing to offer sufficient guarantees to Belgian society about the quality of the diplomas conferred by these private universities, was by interfering in the educational and examination programmes that resulted in those diplomas. Of course, this necessitated a direct attack on the highly appreciated German-style internal freedom of the universities, and contributed to the fact that also this specific characteristic to a certain degree petered out into a rhetoric dream.

German Supremacy and How to Deal with It

From the 1860s, German universities were increasingly referred to as sources of inspiration for the reforms needed at their Belgian counterparts. Certainly their great intellectual freedom contrasted to a large extent with the authoritarian attitude of the Belgian government, according to a great number of Belgian professors and students. The lack of freedom at Belgian universities hampered the scholarly attitudes of both the students and the professors, which in turn impeded the introduction of greater freedom, because they would be incapable of dealing with it. For an increasing number of professors, the German universities seemed to be the polar opposites of their own.[3] A simple comparison of their educational programmes revealed a great deal, it was said. As in France, the Belgian academic and political authorities were prone to including as many courses as possible in each programme, resulting obviously in a loss of quality. Most of the courses were extremely superficial, as many professors complained. The German universities, on the other hand, chose to offer narrower but deeper programmes of education and, especially, examinations, so that the lecturers could aim for more in-depth knowledge, according to Auguste Wagener and Joseph Gantrelle, two professors at the State University of Ghent.[4]

Many professors who visited German, mainly Prussian, universities tried to identify the reasons for their higher level of scholarly endeavour. Most essential was the fundamental principle that ruled the University of Berlin, according to Joseph Demarteau, a scholarship student in Berlin in 1861 and later professor at the teacher-training institute at the State University of Liège:

> *Cultiver la science pour la science*, without paying attention to possible applications of scientific research in daily life. The University of Berlin aims for a way of doing science without any foreign elements disturbing it, assuming that science can progress by developing itself organically in a spiritual sphere, without any contact with the outside world and without the demands of government or the economic market dictating the course of science. This basic principle dates from Humboldt and Altenstein, and is often repeated at academic ceremonies.[5]

A more explicit reference to the idea of *Einsamkeit und Freiheit* is hardly possible.[6]

As early as 1860, in the republication of his report on higher education in Prussia, a colleague from Liège had pointed to the fundamentally different philosophical background at the university in Berlin and, by extension, at

3 Pieter Dhondt, *Un double compromis: Enjeux et débats relatifs à l'enseignement universitaire en Belgique au XIXe siècle* (Ghent: Academia Press, 2011), 390.

4 Auguste Wagener & Joseph Gantrelle, 'Discussion des conseils de perfectionnement sur la question des jurys pour la collation des grades académiques', *Revue de l'Instruction publique en Belgique*, 19 (1876), 27f.

5 Jospeh Demarteau, 'Les conférences du séminaire philologique de Berlin', *Revue de l'Instruction publique en Belgique*, 4 (1861), 172.

6 See Gert Schubring (ed.), *'Einsamkeit und Freiheit' neu besichtigt: Universitätsreformen und Disziplinenbildung in Preussen als Modell für Wissenschaftspolitik im Europa des 19. Jahrhunderts*, Stuttgart: Franz Steiner, 1991).

German universities in general. 'Let us have a look at the doctrines that dominated at the University of Berlin', Charles Loomans began his eulogy on the German model.

> At the time of the foundation of this institution, a well-known scholar, Guillaume de Humboldt, was leading the department of public education within the government administration. Humboldt himself teaches us that the intention of the government, when founding the University of Berlin, was to answer the hope that the university could encourage the intellectual progress of the nation....In that way, the University of Berlin was predestined to become the metropolis of science and the model for other institutions of higher education....The government...clearly wanted to promote the idea of *science pour la science*.[7]

These two professors from Liège were the only Belgian scholars at the time to explicitly link the German university system with the name of Wilhelm von Humboldt.[8] On the other hand, the spirit that characterized the ideal (non-existent) German university—that one should pay greater attention to pure scholarship and research—was well received in Belgium. Complaints were legion about the lack of a scientific spirit among Belgian students and professors,[9]

7 Charles Loomans, *Rapport sur l'enseignement supérieur en Prusse présenté en mars 1845, à M. Nothomb, Ministre de l'Intérieur* (Brussels: Lesigne, 1860), 23.

8 One need only consider the literature on the mythologization of the German university system to realize how unique they were. Evidently, Humboldt was not known as an educational reformer before 1880, and his writings on (university) education only began to attract interest after their republication on the occasion of the first centenary of the University of Berlin in 1910. Quite where Demarteau and Loomans got their information from is thus unclear. In their works, both of them devoted a chapter to the history of the German universities and Berlin, so most probably they had done some research on the subject. See Rüdiger vom Bruch, 'Friedrich-Wilhelms-Universität Berlin: Vom Modell 'Humboldt' zur Humboldt-Universität 1810–1949', in Alexander Demandt (ed.), *Stätten des Geistes: Grosse Universitäten Europas von der Antike bis zur Gegenwart* (Cologne: Böhlau, 1999), 257–278; Mitchell Ash (ed.), *Mythos Humboldt—Vergangenheit und Zukunft der deutschen Universitäten* (Vienna: Böhlau, 1999); Sylvia Paletschek, 'Die Erfindung der Humboldtschen Universität. Die Konstruktion der deutschen Universitätsidee in der erste Hälfte der 20. Jahrhundert', *Historische Antropologie. Kultur—Gesellschaft—Alltag*, 10/2 (2002), 183–205; Markus Huttner, 'Der Mythos Humboldt auf dem Prüfstand: Neue Studien zu Wirklichkeit und Wirkkraft des (preußisch-)deutschen Universitätsmodells im 19. und 20. Jahrhundert', *Jahrbuch für Universitätsgeschichte*, 7 (2004), 280–285.

9 Joseph Antoine Spring, 'De l'esprit scientifique à notre époque et dans nos universités', *Annales des Universités de Belgique*, 2 (1860–3), 386–396; Jean Hubert Thiry, 'Discours

and gradually everyone concerned came to agree that the universities had to fulfil a double mission: they had to furnish a society of practitioners as well as of scientists. According to some, German universities showed an exaggerated regard for the theoretical approach to education; for others, the practical, utilitarian approach to education was still too dominant at Belgian institutions of higher education.

Another Liège professor, Louis Jean Trasenster, took a much more provocative standpoint in his explanation of German scholarly superiority. Rather than a handful of specific features of the German university system, he attributed it to other causes, such as the greater freedom of students and professors, and the fact that students were better prepared for university. In an anonymous pamphlet of 1873, he considered the second half of the nineteenth century, 'one of these great stages in history in which the centre of civilization displaced itself'. Due to increasing ultramontanism, compounded by the difficult relationship between religion and science, the Latin countries prevented each form of scientific progress, according to Trasenster. The German scientific world, on the other hand, was characterized by the greatest intellectual freedom, and, as he put it, 'this spirit of intellectual independence even penetrated into the universities in the Catholic parts of Germany'.[10]

The origins of German supremacy, according to Trasenster, should be sought in the fact that after the crushing defeat of Jena in 1806, Prussia 'understood that it had to regenerate itself, and it tried to do so particularly through education. It is principally education at the universities that transformed and saved Germany'.[11] His colleague at the faculty of medicine in Liège, Léon Fredericq, echoed this view some years later. In his eulogy on the physiological education at the University of Berlin, he repeated the legendary, prophetic words of Friedrich Wilhelm III on the occasion of the foundation of the new university in 1810: 'Der Staat muss durch geistige Kräfte ersetzen, was er an physischen verloren hat' ('The state must replace by intellectual powers what it has lost in material power').[12] Sixty years later the prophecy would be fulfilled: in 1871 German science defeated the French, just as the German armies had defeated

d'ouverture prononcé en séance publique le 12 octobre 1874: L'esprit scientifique et la liberté d'enseignement', *Université libre de Bruxelles: Rapport sur l'Année académique* (1873–4), 19–28; and Nicolas du Moulin, *L'esprit scientifique dans les universités: Discours pour l'ouverture solennelle des cours, 1880–1881* (Ghent: Annoot-Braeckman, 1880).

10 Louis Jean Trasenster, *De l'enseignement supérieur en Belgique* (Luik: Desoer, 1873), 12f.

11 Trasenster 1873, 12.

12 Léon Fredericq, 'L'enseignement de la physiologie à l'université de Berlin', *Revue de Belgique*, 3/38 (1881), 119.

the French, something the professors in Liège and many of their colleagues at home and abroad were not slow to acknowledge.

Trasenster was certainly not alone in his analysis of the problem. Many of those involved agreed that the lack of freedom was the main cause for the shortcomings of Belgian university education. That said, not everyone credit this to an increasing orientation towards authoritarian Rome; in fact, most of them searched for an explanation in the typically Belgian contrast between public and private education. Indeed, Catholics and liberals had made use of the freedom of education granted by the Constitution of 1831 to establish their own universities (the Catholic University of Leuven and the Free University of Brussels respectively), but because the government wanted to guarantee the standard of the diplomas conferred by these institutions, their freedom to decide on their educational and examination programmes was extremely limited. Doubly so for the University of Leuven, in fact, for in addition to the supervision of the government, the Belgian bishops had a close eye on the daily business of the university. However, when in 1878 the more progressive Pope Leo XIII succeeded the extremely conservative Pius IX, Trasenster moderated his opinion too.

Moreover, the majority of professors and politicians were convinced that Belgium should not renounce its (Catholic) Latin and French past just like that. On the contrary, they were convinced the country should make better use of its central position between what were seen as practical France and philosophical Germany—'by combining in our work the honesty and accuracy of France and the analytical and philosophical genius of Germany', as it was put by Emile Banning, when looking back on his studies in Berlin.[13] The Belgian universities had to search for the golden mean between the (French) applied–practical and the (German) fundamental–scientific approaches to education. One of the few to do so, Trasenster even added Britain to the picture:

> Belgium has to make use of its location at the crossing of three great nations and three great axes of Western civilization. It has to look for the qualities that distinguish these nations from one another and try to reunite, in a wise kind of eclecticism, the clearness, the precision, and the ability to explain something from the French people; the rationality and tendency to take initiative from the English people; and the persevering

13 Emile Banning, 'Rapport sur l'organisation et l'enseignement de l'université de Berlin', *Annales des Universités de Belgique*, 2 (1860–3), 120.

> self-sacrifice with which the Germans devote themselves to their scientific work and to more abstract studies.[14]

Actual developments on the ground do not entirely tally with the gloomy picture of Belgian university education given in numerous publications and speeches by professors, politicians, and journalists at the end of the nineteenth century. To begin with, many lecturers did take the initiative to offer optional subjects; indeed, in Liège a number of them even entered into competition with the regular professors by giving the same courses. A large number of practical exercises were organized. With the special postgraduate degree that was introduced in 1853, students had the opportunity to specialize and continue their postgraduate studies.[15] In Brussels, students who intended to apply for a position as a lecturer had to write a dissertation; similarly, in order to obtain certain scientific degrees in Leuven, dissertations were also compulsory. Nevertheless, these changes had only very limited effects on the educational and examination programmes of regular students. Following the laws of 1890 and 1891, additional practical, scientific exercises were made obligatory in medicine, but the emphasis was still on clinical exercises and other vocational subjects. A dissertation was made compulsory, but only for Doctor's degree students in arts and sciences. And moreover, the Candidate's degree programmes offered by these faculties remained very important for the general education of all university students.

This ambiguous attitude was fuelled by arguments over content. Throughout the nineteenth century, Belgian professors and politicians struggled to find

14 Louis Jean Trasenster, 'Discours prononcé le 12 octobre 1880, dans la séance d'ouverture solennelle des cours de l'université de Liège (Du rôle de l'enseignement supérieur et des améliorations et compléments qu'il réclame en Belgique)', in Jean Joseph Thonissen (ed.), *Situation de l'enseignement supérieur donné aux frais de l'état: Rapport triennal. Années 1880, 1881 et 1882* (Brussel: Gobbaerts, 1886), 104.

15 The doctorate was introduced as a scholarly degree in the early nineteenth century, with the abolishment of all the (social) privileges connected to the old degree. It gradually took on a more practical bent with the advent of the doctoral dissertation. Some German universities introduced the dissertation as early as 1831, Norway followed in 1847, France at the end of the 1860s, and the Netherlands in 1876—but Great Britain, for instance, only in 1919. However, true comparisons are extremely difficult, due to the very diverse nature of the qualification. See, for example, Alexander Busch, *Die Geschichte des Privatdozenten: Eine soziologische Studie zur grossbetrieblichen Entwicklung der deutschen Universitäten* (Stuttgart: Enke, 1959); and Renate Simpson, *How the PhD Came to Britain: A Century of Struggle for Postgraduate Education* (Guildford: Society for Research into Higher Education, 1983).

a compromise between what they considered the French, broad-based, vocational model on the one hand, and the German system with its focus on science on the other. At least as important, though, was the necessity of striking a balance between public and private universities, and between the unrestricted freedom of education and complete state supervision. The lack of supervision in the private universities would force the government to prescribe an examination programme, it was concluded. This goes far in explaining why Belgium's universities lagged behind in the modernization of its education, compared with the situation in neighbouring countries. In general, new ideas were adopted relatively quickly, but it appeared to be extremely difficult to put them into practice without disturbing the balance between public and private universities. In the event, the unique and unrestricted extrinsic freedom of education led not to greater choice, but only to limits on the intrinsic freedom of the universities.[16]

The Von Bissing University: From Rhetoric to Reality?

The rhetoric of the introduction of the German university system in Belgium received a new, strong impulse during the First World War. The German invasion of Belgium in August 1914 led to the immediate closure of all the Belgian universities, but by September 1915 the German government, under the direction of Moritz Ferdinand von Bissing, was mooting the idea of reopening the University of Ghent, under close supervision, in October of the following year. Walther von Dyck, a professor in Munich, took the lead in the organizational preparations. The new Dutch-speaking university would become the basis of Germany's *Flamenpolitik*—its Flemish policy.[17]

The first difficulty was to attract enough professors for the new university. Only a few of the active professors in Ghent were willing to start teaching at the new institution in October 1916. Therefore, von Dyck was forced to look for others, 'who have followed a clear scientific career, who speak Flemish, and who had some kind of affiliation with Germany *before* the war (!). For instance, they have studied with German scientists, they have published articles or books in German, or they are members of German scientific societies'.[18] It is

16 Dhondt 2011, 403.

17 For the University of Ghent during the First World War, see Marie Bourke, *The University of Ghent 1916–1918* (Ghent: n.p., 1980).

18 'Beilage 2 zum vierten Bericht über die Vlamisierung der Universität Gent', in Elienne Langendries (ed.), *De 'Vlaamsche Hoogeschool' te Gent (1916–1918): Een bronnenuitgave* (Uit het Verleden van de RUG 19; Ghent: Archief RUG, 1985), ii. 551.

not really surprising that many of the possible candidates had studied at the University of Berlin or another German university. Some of them answered the call with enthusiasm; others, such as the historians Paul Fredericq and Henri Pirenne, did not collaborate at all, even to the extent that they were transported to Germany in March 1916.[19]

In addition to the lack of staff, there were a number of other practical problems that had to be solved before the university could reopen. Actually, the entire university system needed to be switched over to the German model, the most crucial point of it, according to von Dyck, being the introduction of greater

> freedom of academic education, which is really the power and pride of our German universities [and which] is unknown in Belgium. There, it is still customary only to take the required courses, which suppresses every chance of a wider interest. The years are not regarded as years of study, but as years of examination. Each year the students have to pass examinations, be it for the Candidate's or Doctor's degrees, and each time these examinations are limited to what is discussed during lectures. It will be clear that a gradual introduction of greater freedom for students and professors is absolutely crucial for the development of the university; moreover, in that way it will address something that has long been a source of complaint among Belgian professors.[20]

Although some optional courses were introduced between October 1916 and October 1918, giving practical expression to this plea for more freedom in education, the examination system was not thoroughly reformed, and so it remained difficult to attract enough students to these courses. Ultimately, since the Von Bissing University, as it was called, only existed for two years, many of the principles and ideals it was said to stand for were never realized.

In October 1918, the Germans fled Ghent, and within two months the city's university reopened as a French-speaking institution led by Fredericq, who in the spring of 1919 was succeeded by the other national historian, Pirenne. Pirenne would later devote his farewell address in 1921 to the topic 'Ce que

19 Henri Pirenne, *Souvenirs de captivité en Allemagne (Mars 1916–Novembre 1918)* (Brussels: Maurice Lamertin, 1920); and Kristoffer Nyrop, *The imprisonment of the Ghent professors: a question of might and right: My reply to the German legation in Stockholm. With appendixes* (London: Hodder & Stoughton, 1917).

20 Walther von Dyck, 'Die Umwandlung der Genter Universität in eine flämische Hochschule', in Langendries 1985, ii, 727–728.

nous devons désapprendre de l'Allemagne' ('What we have to unlearn from Germany'). In this speech, he admitted that he had once admired German science, just like most of the other scholars of his generation: 'No one who lived in that period could escape from it. Everyone moved to Bonn, Leipzig, Munich, Berlin to learn good practices and returned very enthusiastic. Once they returned to their own country, these young people tried to introduce the merits of the school they had attended at their own institution. Their language was full of German words. They were ever talking about *Excursen*, *Antiquariats*, *Regesten*, and *Forschungen*.[21] Some of them succeeded in introducing a couple of new ideas and methods, such as *cours pratiques* and seminars, inspired by the German example. However, their confidence in their own superiority led these Germanized scholars to ruin, according to Pirenne. He pleaded for 'another view on humanity, according to which all people should not be appreciated for their own value, but only for the value they have for humanity as a whole',[22] and in that respect, in Pirenne's opinion, the Germans had not made a good impression, to say the least.

The Mythologization of Humboldt in Recent Years

The impact of the First World War on Belgium was long-lasting. German scholars faced immense difficulties in rebuilding their credentials again after the war. For instance, the world-famous Solvay conferences in Brussels, which assembled leading physicists from all over the world, explicitly excluded Germans until 1927 (Germany having become a member of the League of Nations in 1926).[23] Nevertheless, Belgian admiration for German science never really waned. Many professors continued to devote themselves to the introduction of the specific characteristics or special chairs that they had become

21 Henri Pirenne, 'Ce que nous devons désapprendre de l'Allemagne', in *Ouverture solennelle des cours et remise du rectorat de l'Université de Gand* (Ghent: Annoot-Braeckman 1921–2), 10.

22 Pirenne 1921–2, 21; see Heinrich Sproemberg, 'Pirenne und die deutsche Geschichtswissenschaft', in Manfred Unger (ed.), *Mittelalter und demokratische Geschichtsschreibung: Ausgewählte Abhandlungen* (Berlin: Akademie, 1971), 375–446.

23 Dirk van Delft, 'Koude drukte. Het laboratorium van Heike Kamerlingh Onnes als internationaal centrum van lagetemperaturenonderzoek', in Leen J. Dorsman & Peter Jan Knegtmans (eds.), *Over de grens: Internationale contacten aan Nederlandse universiteiten sedert 1876* (Hilversum: Verloren, 2009), 42ff.

acquainted with during their studies in Germany before the war.[24] However, when it came to rhetoric, the idea of restoring intellectual relationships with France was revived, and, particularly after the Second World War, the Anglo-Saxon model increasingly became the main point of reference.

Of course, German universities never left the scene completely, but even so the late increase in references to the German university model is striking, as is the bandying about of the name of Wilhelm von Humboldt. Historians' recent demythologization of the Humboldtian model has clearly not reached the university authorities in Belgium yet. In rectorial addresses and other official speeches, and unembarrassed by any shade of nuance, the Humboldtian German university is used as a topos to contrast with the looming commodification of university education.[25] The rhetorical liberties that the current university rectors take with Humboldt's ideas in order to show their implacable, yet often rather symbolic, resistance to anything that smacks of the university as a commercial enterprise, is plain to see.[26] Just two typical examples will suffice.

In his reflections on the history of European universities at the opening of the academic year of 2007–2008, Benjamin Van Camp, rector of the Dutch-speaking Free University of Brussels (VUB), gave an almost mythic air to the foundation of the University of Berlin by Humboldt:

> The university receives complete autonomy with regard to spending its means, and succeeds in combining this freedom with excellence through competition. All university education is based on the solid ground of pure science. The professor is both a researcher and a teacher. He passes on the process of the development of new knowledge to his students'. According to the rector, the success of the Humboldtian model is proven by the fact that 'this university model is copied in all countries, because it

---

24 Sofie Onghena, 'The Survival of 19th Century Scientific Optimism: The Public Discourse on Science in Belgium in the Aftermath of the Great War (ca. 1919–1930)', *Centaurus: An International Journal of the History of Science and its Cultural Aspects*, 53/4 (2011), 280–305.

25 See Christian Krijnen, Chris Lorenz & Joachim Umlauf (eds.), *Wahrheit oder Gewinn? Über die Ökonomisierung von Universität und Wissenschaft* (Würzburg: Königshausen & Neumann, 2011).

26 Anne Rohstock has shown that Belgium is not exceptional in this regard, see Anne Rohstock, 'Some Things Never Change: The Invention of Humboldt in Western Higher Education Systems', in Pauli Siljander, Ari Kivelä & Ari Sutinen (eds.), *Theories of Bildung and Growth: Connections and Controversies Between Continental Educational Thinking and American Pragmatism* (Rotterdam: Sense Publishers, 2012), 165–182.

> forms the driving force of scientific discoveries, and because it pushes forward social progress. Many new universities were established after the example of the Humboldtian model. Our mother university, the Université libre de Bruxelles, was founded in 1834, and, her education being based on free research, she can definitely serve as a good example of the Humboldtian model.[27]

However, recent research has shown, firstly, that the German model, which has indeed exerted a certain attraction within Europe, in fact resists all identification with Humboldt's ideals per se, and, secondly, that in all these cases of intellectual transfer (and as concerns the University of Brussels too) the features that were inspired by German universities were always used in a pragmatic way and adapted to suit the particularities of the indigenous system (even to the extent that frequently the characteristics that were adopted had little in common with the original ideas from which they derived).[28] Later in his address, Van Camp considered the Humboldtian model the continuation of the fabled 'grove of Academus [...] that well-kept garden where Plato and other scholars open-mindedly and freely taught their pupils, far from everyday cares and interference by the authorities'. Van Camp regrets that this idealized picture has been put under pressure by all kinds of developments in society at large in the last two decades, and he concluded by enjoining his audience that 'it is our duty to keep on fructifying the "grove of Academus" and to make it into a paradise on Earth for higher education'.[29] Still, even today, despite Belgium's uncontested extrinsic freedom of education, the intrinsic freedom of its universities is not as far-reaching as Van Camp and many of his predecessors would have liked.

Similarly, the inaugural address by Mark Waer as rector of the Catholic University of Leuven in 2009 uses the Humboldtian university, where 'personal, cultural, or scientific education' is at the centre, as the antipode of the

---

27 Benjamin Van Camp, 'De tuin van Academos: verloren paradijs?', 25 September 2007, <http://www.vub.ac.be/home/academischeopening/archief/0708/benjaminvancamp.pdf>, accessed 1 December 2011.

28 Among many others, Rudolf Muhs, Johannes Paulmann & Willibald Steinmetz (eds.), *Aneignung und Abwehr: Interkulturellen Transfer zwischen Deutschland und Grossbritannien im 19. Jahrhundert* (Bodenheim: Philo, 1998); Rainer Christoph Schwinges, *Humboldt International: Der Export des deutschen Universitätsmodells im 19. und 20. Jahrhundert* (Basel: Schwabe, 2001); Marc Schalenberg, *Humboldt auf Reisen? Die Rezeption des 'deutschen Universitätsmodells' in den französischen und britischen Reformdiskursen (1810–1870)* (Basel: Schwabe, 2002); and Dhondt 2011.

29 Van Camp 2007.

entrepreneurial university. However, in contrast to his colleague at Brussels, Waer was more favourably inclined towards the latter. In his view, a modern university 'has to know how to interpret the ideas of the "entrepreneurial university" in such a way as to strive for the realization of a maximum surplus value, yet without losing out of sight the basic functions of education and pure science', and as for those basic functions, the Humboldtian research university can very well serve as a source of inspiration.[30] The great disadvantage of this Humboldtian university, however, is its very limited interaction with society, at least in Waer's eyes. To a certain extent this picture might be correct for the humanities, but it does not hold good for the way the sciences were approached at German research universities at the end of the nineteenth century, which is what Waer was actually referring to.

Just as in Van Camp's address, Waer uses the name of Humboldt purely to build up a specific argument. Apparently historians are expected to turn a blind eye to how the historical truth is stretched under such circumstances (for example, by unjustly identifying Humboldt's ideals with those of the German research universities at the end of the nineteenth century), and Waer and Van Camp are just two examples of the rhetorical freedoms taken. The extent to which Humboldt's name is tossed around, without really knowing what his ideas actually were, is also shown by the fact that Waer in his speech mentioned 'the report to the minister on the occasion of the foundation of the University of Berlin' written by 'Alexander von Humboldt', instead of his brother Wilhelm.[31] When I mentioned this to him a couple of days after he delivered his speech, the text on the university's website was simply adjusted; meanwhile, the original version has vanished into the ether.[32] Whatever the case may be, even without a detailed analysis of current university policy it is safe to say that rhetoric and praxis differ wildly in this respect. Humboldt's ideas, like those of his British counterpart John Henry Newman, are often used to oppose the current neo-liberal commodification of universities and science, but without any serious impact or even intentions. The name of Humboldt (and Newman) is shamelessly used as a mere rhetorical device.

---

30 Mark Waer, 'De triple helix universiteit: een veranderende universiteit in een veranderende samenleving', 2009, <http://www.kuleuven.be/openingacademiejaar/0910/toespraken/rector.html>, accessed 1 December 2011.

31 Ibid.

32 In all fairness I have to add that it is unclear how or when the mistake was made. Indeed, some of my colleagues who were present at the opening ceremony were extremely impressed by the fact that Waer delivered his speech (of more than 4000 words) without notes, and they were fairly certain the rector had spoken of von Humboldt by surname only.

By placing these figures in their historical context and considering the ways they were viewed in the past, the current generation of professional historians can help nuance the common belief that 'things will never be as good again'. As the rector of the University of Antwerp, Alain Verschoren, put it in October 2010, one cannot just transplant Humboldtian ideas into the current university landscape, yet 'one of the biggest challenges for Europe (and Flanders) in coming years will be to find a good balance between the old [in his eyes] elitist values of Von Humboldt and the changing demands of a mass university with students from extremely diverse social backgrounds and with different levels of education'.[33] Such a contextualization by professional historians can do much to confirm that even today it is still possible, necessary even, to give a historically correct and relevant interpretation of the ideas the university thinkers of the past, and, more generally, of the original background and ideals of the university as a universal institution.[34]

---

33 Alain Verschoren, 'Opening academisch jaar 2010–2011. Toespraak', 2010, <http://www.ua.ac.be/main.aspx?c=.AUHA&n=73306>, accessed 1 December 2011.

34 Pieter Dhondt & Nancy Vansieleghem, 'The Idea of a University: A universal Institution in a Globalised world', *History of Education & Childrens Literature*, (forthcoming 2014).

CHAPTER 6

The Regeneration of the University

Karl Jaspers and the Humboldtian Tradition in the Wake of the Second World War

Johan Östling

The Return of the University

Universities were among the first institutions that the Occupying Powers got up and running following the capitulation of Nazi Germany in May 1945. But much had changed. In a very real sense, the outcome of the Second World War resulted in a ground shift in the academic landscape. As a consequence of the loss of the eastern territories, well-established universities such as Königsberg and Breslau ceased to be German. In the Soviet Occupation Zone, the immediate, forced remodelling of the universities that ensued had clear ideological overtones. Famed German seats of learning such as Jena, Leipzig, Halle, and Berlin were to spend forty years under Communist jurisdiction.[1]

In West Germany, meanwhile, much of the old system lived on. Nazism, the war, defeat, and occupation did not reverse the fundamental order that had been established in the nineteenth century. The organization, the faculty structure, the various subjects' standing: in all important respects, the structure remained the same. None the less, even in the western Occupation Zones there were a number of serious challenges to be faced. Several of the smaller university cities such as Marburg, Göttingen, and Tübingen had been largely spared from physical destruction, but in many of the larger cities such as Berlin, Cologne, and Hamburg, university lecture halls, libraries, and laboratories had been badly damaged.[2] Another significant problem was the question of who was going to populate the universities. When the Nazis had come to

1 Norman M. Naimark, *The Russians in Germany: A History of the Soviet Zone of Occupation, 1945–1949* (Cambridge, Mass.: Belknap, 1995), 440–448; John Connelly, 'Humboldt Coopted: East German Universities, 1945–1989', in Mitchell G. Ash (ed.), *German Universities Past and Future: Crisis or Renewal?* (Providence, RI: Berghahn, 1997).

2 Barbara Wolbring, *Trümmerfeld der bürgerlichen Welt: Universität in den gesellschaftlichen Reformdiskursen der westlichen Besatzungszonen (1945–1949)* (Göttingen: Vandenhoeck & Ruprecht, 2014), 4; Walter Rüegg (ed.) *A History of the University in Europe: Universities since 1945* (Cambridge: CUP, 2011).

 | DOI 10.1163/9789004271944_008

power, many of Germany's finest minds had been forced to flee, not least Jewish scholars. This bloodletting was to have far-reaching consequences for Germany's post-war intellectual life. In addition, a great many of the remaining professors and lecturers were politically compromised, and not a few had been ideological enthusiasts who had gone to some lengths to support of the Nazi cause. Academic denazification was easier said than done, and in many places incomplete.[3]

As the work of getting the universities operational again got underway, the immediate post-war period saw an intensive debate on the idea of the university. To some extent these discussions were part of the profound soul-searching over the German legacy that followed Nazism's fall, but they also drew on a long German tradition of deliberation on the university's mission and character. A central question in the wake of the Second World War was how best to breathe life into an intellectually and scientifically maimed Germany. What role did the Humboldtian tradition and interpretations of the classical German legacy play when the moment came to regenerate the universities after Nazism? In what way was Wilhelm von Humboldt a lodestar in the debate? In this chapter I will look for answers to these questions.

The Humboldtian Tradition Reconsidered

In the past two decades, historians such as Mitchell Ash, Rüdiger vom Bruch, Sylvia Paletschek, and Walter Rüegg have thoroughly historicized the Humboldtian tradition. Their thesis, taken back to first principles, could be said to be that Humboldt was born not in 1810, but at the earliest in 1910. The basis for this, their dramatic formulation, was the realization that Wilhelm von Humboldt was little in evidence in the nineteenth-century debates about the university: his writings were unknown; his ideas had little impact; and it was the Enlightenment universities of Halle and Göttingen, not their Prussian equivalent in Berlin, that were the yardsticks. However, at the turn of the twentieth century Humboldt was suddenly discovered—in Prussia, at least. His treatise on education was published and became famous when the University of Berlin celebrated its centenary in 1910. At much the same time, influential

3 A number of studies have looked at the universities' denazification: for the literature, see Jan Eckel, *Geist der Zeit: Deutsche Geisteswissenschaften seit 1870* (Göttingen: Vandenhoeck & Ruprecht, 2008), 88–111; Axel Schildt & Detlef Siegfried, *Deutsche Kulturgeschichte: Die Bundesrepublik—1945 bis zur Gegenwart* (Munich: Carl Hanser Verlag, 2009), 54–57; and Wolbring 2014, 19–22.

educationalists and teachers such as Adolf von Harnack and Eduard Spranger gave new currency to his ideas.[4]

The lion's share of the recent research on Humboldt dwells on the decades around 1900; however, such analyses as there are of the 1920s and 1930s suggest that he remained a force to be reckoned with. In the Weimar Republic, the neo-humanist foundations he represented were thought to guarantee stability in a turbulent age; during the Third Reich he became a hated figure whom it was necessary to vilify. All this serves to underline a central insight: what had once appeared a *supra*-historical ideal was in fact mired in the complexities of successive epochs, and inspired by turns criticism, reform, and the glorification of past greatness.[5]

There are very few studies of the Humboldtian tradition in the post-war period.[6] In a survey of the West German university system, however, Konrad H. Jarausch argues that 'the postwar chaos prompted a return to *Humboldtian rhetoric* as an uncompromised tradition', even though this once inspiring vision 'had rigidified into a ruling discourse'. By reviving the rhetoric of *Bildung*,

4 See, for example, Walter Rüegg, 'Der Mythos der Humboldtschen Universität', in Mathias Krieg & Martin Rose (eds.), *Universitas in theologia—theologia in universitate: Festschrift für Hans Heinrich Schmid* (Zurich: Theologischer Verlag, 1997); Rüdiger vom Bruch, 'A Slow Farewell to Humboldt? Stages in the History of German Universities, 1810–1945', in Mitchell G. Ash (ed.), *German Universities Past and Future: Crisis or Renewal?* (Providence, RI: Berghahn, 1997); Sylvia Paletschek, *Die permanente Erfindung einer Tradition: Die Universität Tübingen im Kaiserreich und in der Weimarer Republik* (Stuttgart: F. Steiner, 2001a); Sylvia Paletschek, 'Verbreitete sich ein "Humboldt'sches Modell" an den deutschen Universitäten im 19. Jahrhundert?', in Rainer Christoph Schwinges (ed.), *Humboldt international: Der Export des deutschen Universitätsmodells im 19. und 20. Jahrhundert* (Basel: Schwabe, 2001b); Sylvia Paletschek, 'The Invention of Humboldt and the Impact of National Socialism: The German University Idea in the First Half of the Twentieth Century', in Margit Szöllösi-Janze (ed.), *Science in the Third Reich* (Oxford: Berg 2001c); Marc Schalenberg, *Humboldt auf Reisen? Die Rezeption des 'deutschen Universitätsmodells' in den französischen und britischen Reformdiskursen (1810–1870)* (Basel: Schwabe, 2002); and Thorsten Nybom, 'Humboldts Vermächtnis: Betrachtungen zu Vergangenheit, Gegenwart und Zukunft des europäischen Hochschulwesens', in Bernd Henningsen (ed.), *Humboldts Zukunft: Das Projekt Reformuniversität* (Berlin: Berliner Wissenschafts-Verlag, 2007). This research is discussed in Johan Östling, 'The Humboldtian Tradition: The German University Transformed, 1800–1945', in Pieter Dhondt (ed.), *University Jubilees and University History at the Beginning of the 21st Century* (forthcoming); and Martin Eichler, 'Die Wahrheit des Mythos Humboldt', *Historische Zeitschrift* (2012), 294.

5 Paletschek 2001b; Östling (forthcoming).

6 See, for example, Rüegg 2011.

Jarausch continues, 'the universities also returned to the problems of elitism, arrogance, and apoliticism'. One effect of this was that the West German universities were to be haunted by 'Humboldt syndrome' in the post-war period, something that made much-needed democratization and reform more difficult.[7] Dieter Langewiesche is not as inclined to hail the return of Humboldtian rhetoric. He has studied 142 inaugural addresses by German university rectors in the period 1945–1950, and found that Humboldt was mentioned on only seven occasions. Moreover, it was principally in Berlin that the name of the Prussian educational reformer was invoked; in the rest of the German-speaking area they tended to refer to a German university model instead.[8]

Without doubt the most important study in the field is Barbara Wolbring's solid Habilitation thesis on the immediate post-war debate about Germany's universities in the Western occupation zones. She makes clear just how rich and varied the discussion about the universities' role and the meaning of education was in these years. In so doing, she firmly refutes the notion common in earlier research that the occupation years were notable for the vacuum in scholarly self-understanding. It is true that many of the ideas that had flourished in the previous half century had lost their attraction along with their legitimacy, but proposals for what the future ought to be were numerous. One group can be said to have worked for something amounting to a Humboldt revival, but the specific ideals they associated with his name and the exact manner in which they evoked the older German tradition do not feature in Wolbring's analyses.[9]

A central figure in the German university debate was the philosopher Karl Jaspers. In a series of speeches, articles, and books he expounded on how the universities would best be regenerated and what shape the ideal of the university should take after the great catastrophe. In an exchange with the Romanicist Ernst Robert Curtius, he was even referred to as 'a Wilhelm von Humboldt of our time'.[10] His impact on the debate was all the greater because

7 Konrad H. Jarausch, 'The Humboldt Syndrome: West German Universities, 1945–1989', in Mitchell G. Ash (ed.), *German Universities Past and Future: Crisis or Renewal?* (Providence, RI: Berghahn, 1997), 35–38.

8 Dieter Langewiesche, 'Das deutsche Universitätsmodell und die Berliner Universität', in Ilka Thom & Kirsten Weining (eds.), *Mittendrin: Eine Universität macht Geschichte* (Berlin: Akademie Verlag, 2010), 26; Dieter Langewiesche, 'Humboldt als Leitbild? Die deutsche Universität in den Berliner Rektoratsreden seit dem 19. Jahrhundert', *Jahrbuch für Universitätsgeschichte*, 14 (2011).

9 Wolbring 2014.

10 Quoted in Mark W. Clark, *Beyond Catastrophe: German Intellectuals and Cultural Renewal after World War II, 1945–1955* (Lanham, Md.: Lexington Books, 2006), 73.

of his enduring position as a public moral authority in a country that, more than any other, needed to examine its conscience, and badly needed direction. It is for this reason that I will therefore concentrate on Jaspers's idea of the German university.

Karl Jaspers and the Rebirth of the German University

Karl Jaspers was born in Oldenburg in 1883 and grew up in a wealthy banking family. After a couple of terms studying law at university, he switched to medicine, and in 1908 took his doctorate in Heidelberg with a thesis on psychiatry. Encouraged by Max Weber and Wilhelm Windelband, he gravitated towards psychology, a subject at that time reckoned to be part of philosophy, and in 1913 published the epoch-making *Allgemeine Psychopathologie*. Jaspers's scholarly peregrination culminated in the early 1920s, when he left his chair in psychology for one in philosophy. During the inter-war years he published the series of major works that set out the principles of his existential philosophy.[11]

Karl Jaspers was long the very image of the apolitical mandarin. The rapid headway made by the Nazis did not perturb him; even after the *Machtergreifung* he was content to dismiss National Socialism as an operetta. Yet equally, it was not long before it dawned on him just how unpopular he was with Germany's new masters. In 1933 he was removed from his positions of trust; in 1937 he was forced to take early retirement; in 1943 came a publication ban on his works. Jaspers's crime was not that he had spoken out against the regime—it was that he had refused to annul his marriage with his Jewish wife.[12]

The Jaspers spent the war in Heidelberg under the constant threat of deportation. Several times they only managed to avoid deportation at the very last moment. Yet despite pressing anxiety and intellectual quarantine these difficult years were as much a respite for Jaspers, who continued his philosophical work, dedicating himself to his contemplation of his own and his country's development. It was from his own experiences that he drew his ideas about what caused the German catastrophe and how the Germans might once again rise from its ashes. A month or so before the end of the war he wrote in his

---

11 Suzanne Kirkbright, *Karl Jaspers: A Biography: Navigations in Truth* (New Haven: Yale University Press, 2004); Hans Saner, *Karl Jaspers: Mit Selbstzeugnissen und Bilddokumenten* (Reinbek bei Hamburg: Rowohlt, 2005); Kurt Salamun, *Karl Jaspers* (Würzburg: Königshausen & Neumann, 2006).

12 Kirkbright 2004; Saner 2005.

diary, 'Whoever survives [war's end] will certainly have a task marked out for them that will consume the rest of their lives'.[13]

Jaspers's mission became the restoration of Germany's intellectual honour. He realized that this would call for more than cool observation and abstract thought. As a philosopher, he had to break with the apolitical line of the German tradition and instead formulate a message that would speak to far wider groups and have far greater dramatic impact. He had to shoulder his social responsibility in earnest. An important channel was to be *Die Wandlung* ('The Transformation'), the monthly periodical that Jaspers began in the autumn of 1945 with Dolf Sternberger and others, and whose watchwords were humanism, freedom, and spiritual renewal.[14]

Real resurgence for Germany would depend on a true examination of German history, Jaspers believed. It was not enough to return naively to the pre-Nazi position. It was equally futile to imagine that the Germans could simply break with the past. Instead, what had to be done was to draw up history's balance sheet, a summing-up of the German nation's liabilities and assets. In his controversial book *Die Schuldfrage* (1946; *The Question of German Guilt*, 1947), he tackled the question of where the moral responsibility for the country's crimes lay, and in so doing made one of the most important contributions to Germany's self-examination immediately after the war. However, true rebirth presupposed that all that was good and edifying about the Germany's heritage would be put to good use. The universities had an absolutely vital role to play in this spiritual—and by extension political—renaissance. If Germany was to rise again, people had to be able to put their faith in the country's universities.[15]

In the immediate post-war period, and especially 1945–6, Jaspers returned again and again to the fate of the university. In a large number of speeches, essays, and newspaper articles he presented variations on the same theme: it was of crucial importance to Germany that the university regained its strong position. At the very end of the war he had put the final touches to his *Die Idee der Universität* (1946), the most detailed account of the university's nature and purpose to be published in the early post-war period. In 1923 Jaspers had published a book with the same title, but he now stressed in the foreword that

13 Quoted in Saner, 2005, 51. Unless otherwise noted, all translations from the German are made by Charlotte Merton.

14 Ralf Kadereit, *Karl Jaspers und die Bundesrepublik Deutschland: Politische Gedanken eines Philosophen* (Paderborn: Schöningh, 1999); Clark 2006.

15 Kadereit 1999; Clark 2006; Jennifer M. Kapczynski, *The German Patient: Crisis and Recovery in Postwar Culture* (Ann Arbor: University of Michigan Press, 2008).

the new treatise was 'not a second edition, nor a reworking of the text, but is instead a new conception grounded in the experience of the two terrible proceeding decades'.[16] Barbara Wolbring points out that Jaspers's post-war treatise was to become 'an important reference point for discussion; here the concept was formulated out of our own neo-humanist, idealistic German university tradition, from which a new beginning could be attempted'. He gave voice to a significant current of feeling among the professors of the day.[17]

Die Idee der Universität is an important document if we are to understand Jaspers's general conception of scholarship, education, and the university. In the 123-page book he ranges from the abstract to the specific: he begins with a theoretical consideration of the nature of scholarship, moves on to the university's mission, and concludes by addressing the questions of politics and economics. If one is to understand Jaspers's vision of how the university could be regenerated, and how this rebirth was bound up with one particular interpretation of German history, we have to look at other texts from the same period, above all some of his speeches that were later published.[18]

The Vision of Regeneration

Three months after Germany's capitulation, in August 1945, Jaspers gave a speech on the occasion of Karl Heinrich Bauer's installation as the newly elected rector of the University of Heidelberg and the resumption of teaching there. It would later be published *in extenso* in the first issue of *Die Wandlung* under the title 'Die Erneuerung der Universität'. Rather than 'revival', the word *Erneuerung* should be understood as 'regeneration'—an infusion of new life into the university akin to a rebirth or renaissance.[19]

In his address, Jaspers elaborated on his vision of how the university was to be regenerated. From the very outset, the tone he adopted showed he was sensible to the fact it was a momentous day for the university. After twelve years, it had finally been possible to elect their own rector. What they were now witnessing was a new beginning. Despite all that had happened, the essence of the university had remained untarnished; despite the havoc wrought by the

16 Karl Jaspers, *Die Idee der Universität* (Berlin & Heidelberg: Springer-Verlag, 1946a), 5.

17 Wolbring 2014, 425.

18 The majority of his speeches and texts are to be found in Karl Jaspers, *Erneuerung der Universität: Reden und Schriften 1945/46*, ed. Renato de Rosa (Heidelberg: Verlag Lambert Schneider, 1986).

19 Jaspers 1986, 293.

Nazis, despite their interference in research and teaching, there were professors and students who had made a stand. And since the scholarly spirit had not been crushed, the university could now be resurrected. The people they had to thank for this, continued Jaspers, were the Occupying Powers. Their forbearance, perhaps in fact their help, was essential in being able to rebuild the university again.[20]

'The rebirth of our university cannot simply involve restoring the conditions found before 1933, however', Jaspers concluded in a telling formulation. Too much had happened; the catastrophe had struck too deep. The Germans had changed since then, he continued, and he went on to remind his audience of what the preceding years had brought by way of death, suffering, and humiliation. But the chain of evil had to be broken. They had look long and hard at their true past to find the foundations on which they could build. They would be true to their parents, their home, and their country, the land of Kant, Goethe, Lessing, and other great figures.[21]

It was in this spirit that Jaspers believed the university would be rebuilt. When it came to its physical form, he nursed the hope that its old self-government might be retained and that all the institutions and departments might re-form. Similarly, he hoped that the students would return, and that research and teaching would resume under the same conditions as before. But none of this would mean renewal in any shape or form. Instead, as he said:

This regeneration in fact can only come about through the individual efforts of researchers and students in the community of their intellectual lives. This community must take as its guide the incorruptible ideal of the university—the idea of a university in which research and teaching work together and in common; where freedom to teach and freedom to study are seen as conditions for responsibly exercised independence, and are expected of every individual teacher and student; where rote schoolwork and exclusionary specialization are rejected, and instead the unity of knowledge is developed through living communication and intellectual struggle.[22]

Jaspers argued that renewal of this kind would be evident in the temper of the seminars, departments, and clinics; it would be evident in their publications and textbooks. And it was they alone who could be responsible that renewal. However, as yet there was no common ground on which to meet. The idea of the university had yet to become truly alive again.[23]

20 Karl Jaspers, 'Erneuerung der Universität', *Die Wandlung*, 1 (1945/1946), 95–96.

21 Ibid. 96f.

22 Ibid. 97f.

23 Ibid. 98.

In the beginning, the university was a single whole. Its three faculties were designed to serve people's basic needs: the theological faculty, the soul's salvation; the law faculty, social order; the medical faculty, the body's wellbeing. In Jaspers's eyes it was an abomination that this unity was abandoned in the second half of the nineteenth century, with untold consequences. On the one hand, knowledge was dispersed and muddied by unscholarly approaches. This reached its nadir in the Nazi period. On the other hand, it led to an inability to tackle the real forces of the age, in particular technology, and to see how they impacted on the whole. Renewal had to spring from the past, partly to broaden the university's mission to the point where it embraced all parts of existence, and partly to win back the scholarly idea of unity. Possibly, a new technological faculty ought to be founded and the philosophical faculty once again forged into a single unit.[24]

There were thus several components to Jaspers's thinking on the regeneration of the university. First and foremost, it was a question of making the most of such scholarly spirit as survived the Nazis' barbarism. At heart the university was untarnished, announced Jaspers, and recalled a metaphor that had been coined in the 1920s by the then education minister, Carl Heinrich Becker: the German university was 'fundamentally sound' (*im Kern gesund*).[25] Second, there was a pressing need to restore unity: the university could not be a mere assembly of vocational colleges and specialisms.[26] In addition, it had to broaden its mission and turn its attention to the issues of the day, not least the role of technology in society.

Last but not least, Jaspers's vision of the *Erneuerung der Universität* was closely related to his understanding of university history. There had to be a return the ideals that underpinned the classical German university, he believed. His historiographical views can be glimpsed in various speeches and texts from the early post-war period. The single most cohesive account came in an address he gave in Heidelberg in January 1946. His was the first in a series of lectures by leading German academics that were intended to contribute to a democratic

24 Ibid. 103.

25 Axel Schildt, 'Im Kern gesund? Die deutschen Hochschulen 1945', in Helmut König, Wolfgang Kuhlmann & Klaus Schwabe, (eds.), *Vertuschte Vergangenheit: Der Fall Schwerte und die NS-Vergangenheit der deutschen Hochschulen* (Munich: Beck, 1997); Christoph Markschies, *Was von Humboldt noch zu lernen ist: Aus Anlass des zweihundertjährigen Geburtstags der preussischen Reformuniversität* (Berlin: BUP, 2010), 18.

26 He enlarged on these and other aspects of his notion of scholarly holism in other works, above all in Jaspers 1946a, 37, 43, and 75f.

renewal. The title of his lecture, 'On the living spirit of the university', witnesses to his view that the tradition lived on.[27]

On the Living Spirit of the University

'The university is ancient', were Jaspers's opening words in his historical survey. It was an institution that had its ideational roots in Ancient Greece. The medieval university had been European, wholly dominated by theology, and had left unparalleled modes of thought for posterity. This old-style university was as remote as it was admirable, said Jaspers.[28]

Jaspers then turned to the German-speaking area, sketching the development of the university from the Reformation onwards, divided into three successive stages. The *Protestant university* took upon itself to satisfy the State's need for clerics and government officers. In comparison to the medieval university, the seats of learning of the sixteenth and seventeenth centuries represented an intellectual retrenchment. Humanism was transformed into a new but poorer scholasticism. 'The atmosphere grew bleak, the horizon narrow', Jaspers said.[29]

During the eighteenth century, the *classic humanist university* emerged. The students liberated themselves and sought out the lecturers who had something to offer. A stratum of German educational bourgeois respectability evolved that was based in the large cities. Scholarship and philosophy joined forces and became as one. Finally, Kant, Fichte, and Schleiermacher sketched out the idea of the university, and Humboldt gave it practical shape, as realized in the foundation of the University in Berlin. Everything turned on the freedom of research and teaching, and the level was determined by the inseparability of the two. Realizing this ideal was certainly no simple matter, but nevertheless it created a scholarly community. Humanism became the cornerstone of education and left its stamp on all students, whether they were to become priests, doctors, judges, teachers, or public officials. Major figures contributed to the university's reputation and enjoyed high public esteem.[30]

---

27 Jaspers 1986, 294–295. The title of the lecture, 'Vom lebendigen Geist der Universität', is a play on a well-known inscription at the University of Heidelberg, 'Dem lebendigen Geist' ('To the living spirit'), a motto coined in the 1920s by the literary historian Friedrich Gundolf.

28 Karl Jaspers, 'Vom lebendigen Geist der Universität', in Karl Jaspers and Fritz Ernst, *Vom Lebendigen Geist der Universität und vom Studieren: Zwei Vorträge* (Heidelberg: Schneider, 1946b), 224f.

29 Ibid. 225.

30 Ibid. 225f.

However, in 1848 this picture changed, Jaspers argued. The March Revolution had originally enjoyed considerable support among university professors, but in the event its outcome was not to the university's benefit. In the wake of the revolution, positivist scholarship and idealess realism were segregated. The glorification of Bismarck triumphed over the old professors' worldly wisdom. At the same time, there was growing criticism of the ossified Prussian universities, put at its sharpest and most eloquent by Nietzsche. It was clear that the most significant thought no longer stemmed from the university.[31]

It was in the middle of the nineteenth century that what Jaspers called the *modern university* emerged. Its advocates proclaimed themselves willing to retain the old ideas, but in dilute form. They developed no principles of their own, and showed that they had forgotten their origins by doubting the value of the unity of the university. Instead, they founded institutes of technology and permitted the old philosophical faculties to be divided in two, the one for the sciences, the other for philosophy. The students poured in, the numbers of professors rose: the mass university took shape. Demands and course requirements were lowered now that not all students had a shared background at the humanist gymnasiums. The teaching became more technical and instrumental. As standards fell, specialization continued to grow: 'The intellectual barbarian could shine by serving with merit as a technical specialist', as Jaspers put it. The status of the independent institutes grew rapidly, but their importance did not grow in proportion to their size.[32]

The result was that specialist knowledge displaced philosophical ideas. No one took responsibility for the holistic view of the university, and instead of a wide-ranging, lively debate, academe was dominated by turf wars and sophistry. 'We professors were no longer research figures as in the days of classical idealism', Jaspers pointed out. The course of a professor's life became increasingly that of a public official: few could really dedicate themselves to shaping scholarship. 'The cerebral aristocracy seemed to have come to an end'.[33]

For each decade that passed, people looked back on the previous one with a sense of loss, Jaspers believed. The decline was everywhere to be seen. When he himself went to university in about 1900 the catalogue of woes was longer than ever. There were still the likes of Mommsen, Dilthey, and Weber. But what would happen when that generation was gone? asked Jaspers in a speech as early as 1929, and he went on to quote a longish passage from what he had said then, when, during the first years of the Great Depression, he had put his mind

31 Ibid. 226f.

32 Ibid. 227f.

33 Ibid. 228.

to an appraisal of German history. The Germans had not generated a State of truly global significance. Neither had they left their mark on any human type, nor created any significant culture. All the things of any lasting worth that had come about in the past four hundred years stemmed from the universities. This was true of the most original contribution—Protestantism—and only in Germany were the great, universal philosophers—Kant, Hegel—at the same time professors. The universities in the nineteenth century came to represent all that was best about the German character. Today it was different. The future for Germany and her universities was uncertain. 'The only thing that is clear that the water is rising and all might yet be inundated', said Jaspers as he concluded his speech in 1929.[34]

Jaspers admitted that the catastrophe, when it came, was of a wholly different compass to what he had divined a few years before the Nazis seized power. For a party did indeed emerge that led to the Holocaust that Germany now beheld. The violence of the political inroads into the universities sent them reeling. Instead of an independent academy, Germany was left with a school that took orders from Berlin. The rectors and deans still existed, but in name only, and they were appointed by the National Socialists. If Hitler's Germany had been victorious, the universities would have been dissolved and replaced by vocational colleges and Nazi Party schools. Putting up almost no resistance, the universities and academics capitulated. Everything they held dear—truth, scholarship, dignity—had been left to individuals to defend, Jaspers reminded his listeners.[35]

The consequence, he continued, was that as a university they lost their dignity in 1933. Only with a new university could they rally once again, but this time under very different circumstances and with a new mission. One thing had been very clear during those twelve years: the people who had been most harshly critical of the spiritual poverty of the modern university were the Nazis. Therefore it could be said that the considerable sympathy shown for their criticism demonstrated the sheer scope of the intellectual bankruptcy of the modern university. Yet this critique was not invented by the Nazis—it merely reproduced the academy's own self-criticism. Jaspers owned that it was true, it had at first appeared as if the Nazis wanted to change the university in a welcome direction: faculties working more closely together; a synthesis of the sciences; high seriousness as the basis for all scholarship; a receptiveness to popular and grand narrative; due respect for the great figures. It all sounded

34 Ibid. 228ff.

35 Ibid. 230.

quite splendid, and many had allowed themselves to be charmed by it. But it came to naught. The Nazis had nothing positive of their own to contribute. Yes, in truth they achieved the diametric opposite of what they had promised. Instead of insight, empty talk; instead of *Weltanschauung*, a mishmash of incoherent sententiæ. It dawned on the intellectual Estate that it was for them the bell tolled, while the unintelligent and spineless kept their eyes on the main chance.[36]

Jaspers finished his historical disquisition by asserting, 'The Nazi rape has ceased. Independent research can once again breathe free. Open disagreements are once again permitted and advisable'.[37] With this, Jaspers turned to the present, and asked what their concerted mission should be. In a crucial passage, he declared:

Because of our faith in the Humboldt era we do not seek a radical new conception for the forms of our institutions; our purpose for them is to create a conservative revolution. However, we also know this well: that we cannot recreate what for us was the Classical Age of the German university. The expectations of society, the state and the individual, the conceptual framework of contemporary opinion, and our knowledge and capabilities have all become quite different.[38]

The university could by all means reassume its traditional duties, said Jaspers. But much had changed, and the old educational world had been rent apart. They could not simply continue as they had before as if nothing had happened. One important undertaking was a picture of German history that drew on its worthy heritage. For that, research was needed; it was not sufficient to re-evaluate the old ideas.[39]

Christoph Markschies has noted that historical notions of the university can be said to fall into one of two models—decadence or progress. The first often has more than a little of the golden age myth about it.[40] This was certainly the case for Jaspers. His ideal, the classic humanist university, gradually took shape over the course of the eighteenth century, but began to be undermined from the middle of the nineteenth, and with Nazism came utter disaster. At the same time, it is obvious that Jaspers was in two minds about this historical progression. On the one hand, he wanted to see a return to the classical German university, breathing new life into its ideals. It constituted some of the best of

36 Ibid. 231f.

37 Ibid. 232.

38 Ibid. 232.

39 Ibid. 232–235.

40 Markschies 2010, 20.

the German tradition. On the other hand, he was alive to the fact that it was not possible to return. Too much had changed.

Karl Jaspers and the Humboldtian Tradition

In Jaspers's vision of regenerating the German university, the Humboldtian tradition played what at first glance was a subordinate role. In *Die Idee der Universität* he mentioned Wilhelm von Humboldt only a handful of occasions, and in his other work references to the Prussian educational reformer were few and far between. On one occasion, when he was discussing the relations between the university and the state, he did cite Humboldt, although without further reference: 'The state must always remain aware that education works far more effectively without interference'. It is worth noting that passages to this effect are to be found in Humboldt's 'Über die innere und äussere Organisation der höheren wissenschaftlichen Anstalten in Berlin' (1809/1810), but this particular wording is not.[41]

For Jaspers, Humboldt seems to have been a familiar name, but not one of the great figures of the German university tradition: Kant and Weber were far more important sources of inspiration, and he referred to them repeatedly. Nor was the foundation of the University of Berlin in 1810 a historical milestone to his mind. He traced the origins of the classic humanist university to the eighteenth century, with a single, coherent idea of the university coined at the turn of the nineteenth century by philosophers such as Kant and Schleiermacher. To Humboldt fell the honour of making this new university a reality. He thus did not feature for Jaspers either a thinker or as a synthesist.

Against this background it is hardly surprising that Jaspers did not find any reason to evoke Humboldt to rhetorical or ideological ends. Neither did contemporary voices in the university debate such as the politician Erwin Stein and the philologist Karl Vossler seem to have been any more inclined to do so.[42] That said, there was no particular reason why Humboldt should have been of interest in the immediate post-war period. There was never any question of anything like the Goethe cult seen in conjunction with the

41 Jaspers, 1946a, 103; cf. Wilhelm von Humboldt, 'Über die innere und äussere Organisation der höheren wissenschaftlichen Anstalten in Berlin', in *Werke in fünf Bänden*, ii: *Schriften zur Politik und zum Bildungswesen*, (ed.) Andreas Flitner & Klaus Giel (Darmstadt: Wissenschaftliche Buchgesellschaft, 2010), 257.

42 Wolbring 2014, 298f and 354–357.

bicentenary of the Bard of Weimar's birth in 1949.[43] This conclusion is far closer to Dieter Langewiesche's observations about the German university rectors' inaugural addresses than Konrad H. Jarausch's hypothesis about Humboldtian syndrome.

In Jaspers's eagerness to regenerate the German university, he nonetheless fell back on ideals that are now thought of as being Humboldt's: many of the principles he associated with the classic humanist university were identical to those that today are labelled Humboldtian.[44]

One ideal that he often returned to, for example, was his belief in the unity of scholarship. Like many of the other voices in the university debate of the Occupation years, he was worried about academics specializing, and he spoke out against the knowledge schism.[45] Ultimately this tied in with the university's educational mission. At several points in *Die Idee der Universität* he argued for a holistic approach to scholarly thought. It is true, he wrote, that students go to university in order to prepare for a profession, but they expect far more than that. University should represent scholarship in its entirety, and from this whole the students would be able to form their own *Weltanschauung*.[46]

Repeatedly, Jaspers spoke up in defence of 'the connection between research and teaching', another ideal that today is often associated with Humboldt. Jaspers objected to those who wanted to draw distinctions and transform the universities into research institutes pure and simple. It could not be denied that a good lecturer was not necessarily a good scholar, but only if research and teaching were united could students come into contact with the real knowledge process.[47]

Academic independence of all shades was another major topic where Jaspers adhered to a classic, German tradition without making much mention of Humboldt. At the very beginning of *Die Idee der Universität* he impressed on readers the importance of the fundamental principles of 'freedom in teaching' (*die Freiheit der Lehre*) and 'freedom of study' (*die Freiheit des Lernens*). He would later return to the question several times: the scholar should himself formulate his task and find a path to the solution; the student has the freedom to choose his own direction in his studies.[48] Closely allied with this aspect of

43 Karl Robert Mandelkow, *Goethe in Deutschland: Rezeptionsgeschichte eines Klassikers: 1919–1982* (Munich: Beck, 1989).

44 See the introduction in this book.

45 Wolbring 2014.

46 Jaspers 1946a, 37, 43, and 47.

47 Ibid. 41 and 47.

48 Ibid. 111f and 277f.

academic independence was the university's relationship with the state. Jaspers acknowledged that the university was dependent on the state, but in the wake of the Second World War it also seemed natural to underline the demand for independence from the state.

Admittedly, to touch briefly on these three ideals—the unity of scholarship, the union of teaching and research, and academic independence—is not sufficient to fully characterize Jaspers's philosophy of the university, yet these examples from his most central treatises illustrate how he was moved by the classic German tradition, and wanted to revitalize it. To adopt hermeneutic terminology, we could say that he attempted to *actualize* the classic humanist legacy, but without invoking the name of Wilhelm von Humboldt.[49] He was conscious that the university to some extent had to adapt to the modern world, but he also enjoined the necessity of strengthening academic independence after the Nazis' violent reign. He also knew that it would take long, hard work to breathe new life into Germany's universities, with little hope of immediate success. Yet it was the only alternative. 'In view of the awful reality we face on this path, may we be endowed with courage and confidence', was Jasper's simple hope in concluding his speech on *Erneuerung der Universität*.[50]

49 See Hans Ruin's contribution in this book.

50 Jaspers 1986, 105.

CHAPTER 7

When Humboldt Met Marx

The 1970s Leftist Student Movement and the Idea of the University in Finland

Marja Jalava

By the early 1960s, it had become obvious that most Western countries were seeing an educational expansion at all levels. This was also the case in Finland, which entered the era of 'mass higher education' during the 1970s.[1] To some extent, the extraordinary growth of university education was due to consumer pressure caused by the post-war baby-boom generation.[2] However, a major explanation was the structural change of society, resulting in a notable overall improvement in the average value of exports and a rise in the general level of labour costs. Hence, higher education came to be perceived as a key investment in the promotion of economic growth.[3] At the same time, the Nordic countries started to use educational policies as a strategy for creating social justice and a reasonable standing of living for all social groups.[4] The upshot was the ideal of a uniform and inclusive educational system, which it was thought would be achieved by means of state-led centralized planning. To borrow Aant Elzinga's definition, the 1960s were dominated by a 'radical bureaucratic rationalism' by reform-minded technocrats.[5]

---

1 According to the sociologist Martin Trow, *Problems in the Transition from Elite to Mass Higher Education* (Berkeley, CA: Carnegie Commission on Higher Education, 1973) the concept of 'mass university' refers to the situation in which university students comprise some 20 to 30 per cent of the relevant age group. See also Sirkka Ahonen, *Yhteinen koulu—tasa-arvoa vai tasapäisyyttä?* (Tampere: Vastapaino, 2003), 107–113; Ari Antikainen, Risto Rinne & Leena Koski, *Kasvatussosiologia* (Helsinki: WSOY, 2006), 90–94.

2 See, for example, Eric Hobsbawm, *The Age of Extremes: A History of the World, 1914–1991* (New York: Vintage Books, 1996), 296.

3 Marja Jalava, *University in the Making of the Welfare State: The 1970s Degree Reform in Finland* (Frankfurt am Main: Peter Lang, 2012), 44–50.

4 Arild Tjeldvoll, 'Quality of Equality? Scandinavian Education towards the Year 2000', in id. (ed.), *Education and the Scandinavian Welfare State in the Year 2000: Equality, Policy, and Reform* (New York: Garland, 1998), 3–23 at 3f.

5 Aant Elzinga 'Universities, Research and the Transformation of the State in Sweden', in Sheldon Rothblatt & Björn Wittrock (eds.), *The European and American University since 1800: Historical and Sociological Essays* (Cambridge: CUP, 1993), 191–233 at 214–217.

 | DOI 10.1163/9789004271944_009

However, the avant-garde status of pro-reform technocrats was soon contested by Leftist student activists, who mostly belonged to the younger cohorts born during or immediately after the Second World War. In their eyes, both the traditional 'oligarchy of professors' and the increased power of the technocrats who were bent on modernizing the universities according to the economic interests had to be resisted.[6] In this chapter, I will focus on the criticism of the university system presented by the Leftist student movement in Finland, and their rediscovery of the Humboldtian *Bildungsuniversität* in the 1970s. I will also briefly discuss the role of Humboldt's university conception as an organizational myth of the modern research university.

Finland arguably makes a good case for an examination of the local–global axis in the development of the idea of the university, since comparative and international dimensions were brought to the fore in the post-war Finnish educational debate. On both sides of the barricades, educational experts and academic activists relied heavily on international comparisons when evaluating the applicability of international knowledge and experiences to the Finnish context.[7] Left-wing student radicalization was essentially about transnational networks of activists, organizations, and institutions, which enabled constant flow of ideas and the dissemination of new concepts and symbolic forms, as well as protest tactics, alternative lifestyles, and countercultures. However, actual protest techniques, forms, and ideologies varied greatly from country to country due to the differences in national opportunities, structures, political cultures, and the lenience of individual governments. Thus, although student activists were interested in presenting transnational solutions for global problems such as capitalism and imperialism, there was also plenty of room for national and regional idiosyncrasies.[8] In the following, the Finnish student movement will therefore be considered as a

6 See, for example, Pekka Aarnio, Mikael Böök, Antti Kasvio, Kari Toikka & Matti Viikari, *Johdatus uuteen yliopistoon: Yliopisto, tiede, kapitalismi* (Helsinki: Kansankulttuuri, 1970), 210f.

7 Jalava 2012, 12f; for the general strategy of the Finnish reformers as 'an eclectic avant-gardism of the educated elite in a peripheral country', see Pauli Kettunen, 'The Power of International Comparison: A Perspective on the Making and Challenging of the Nordic Welfare State', in Niels Finn Christiansen, Klaus Petersen & Per Haave (eds.), *The Nordic Model of Welfare—a Historical Reappraisal* (Copenhagen: Museum Tusculanum Press, 2006), 31–65 at 37f and 55–59.

8 Martin Klimke & Joachim Scharloth, 'Introduction to 1968 in Europe', in eid. (eds.), *1968 in Europe: A History of Protest and Activism, 1956–1977* (New York: Palgrave Macmillan, 2008), 1–9 at 4f.

'glocalized' phenomenon in which common global trends were situated in a local space.[9]

A Leftward Turn in Finnish Academe

From an international perspective, the historical development of the Leftist student movement in Finland displays some idiosyncrasies, because of which it might be even considered somewhat exceptional. Leftist thinking in Finnish academe had long been overshadowed by the 1918 Civil War and the right-wing nationalism that ensued. As a result, unlike all the other Western European capitalist countries with a strong Communist Party (above all, France and Italy), the Left had virtually no place at the Finnish universities before the early 1960s, when the first signs of academic Leftism finally hove into view. While socialism had a minimal impact on the academic community, the political influence exerted by academics over the working-class movement was even smaller.[10]

The transformation of the Leftist student movement from a string of anti-bourgeois, single-issue movements into one with a more definite political stance came at the turn of the 1970s. The impetus came in part from international examples, such as the events of May 1968 in France; however, Finnish radicalization was equally the result of domestic politics. The Leftist student activists had had high hopes of the centre–left Popular Front government established in 1966. In the most optimistic accounts, it was even considered the beginning of a peaceful transition toward socialism in Finland.[11] However, in order to enhance the international competitiveness of the export sector, the new government was forced to hike taxes and prices and conduct a massive 31 percent devaluation of the Finnish mark (*markka*) in 1967. Moreover, as a part of the general stabilization plan, in 1968 the government, trade unions, and employers' organizations hammered out the first national income policy agreement to prevent inflation rising because of feared rise in wages. In the

9 For the interplay between global and local dimensions, see, for example, Roland Robertson, 'Glokalisierung', in Ulrich Beck (ed.), *Perspektiven der Weltgesellschaft* (Frankfurt am Main: Suhrkamp, 1998), 192–220 at 215ff.

10 Risto Alapuro, 'Studentrörelserna och sociologin i 60-talets Finland', *Finsk Tidskrift* 6–7 (1976), 336–360 at 338–342; Veikko Tarvainen, 'ASS ja 30-luvun Soihtu', in Tapio Bergholm (ed.), *Punaisista apostoleista opiskelijaradikalismiin: SONK:n 20-vuotisjuhlakirja* (Helsinki: Suomen Sosialidemokraattisen Opiskelijanuorison Keskusliitto, 1983), 55–71 at 56–69.

11 Kimmo Rentola, 'Nuortaistolaisuuden synty', *Politiikka*, 4 (1990), 243–260, at 255.

eyes of the radical Left, the Leftist parties had fallen into the trap of collaborating with a coalition whose aim was to institute neo-capitalist planning, harnessing various interest groups in the defence of the centralizing zeal of Finnish capitalists.[12]

In the early 1970s, the far-Left Western European student radicals were usually well to the left of the official Communist parties. For instance, in Sweden and Norway, the Marxist–Leninist students adopted an anti-Soviet stance, ending up supporting Maoism.[13] In Finland, meanwhile, the radicalized students verged upon the pro-Soviet, rigidly orthodox, hard-line minority in the official Communist Party, which, despite its obvious dogmatism, was perceived as the only realistic force in Finnish society—comparable to the Soviet Union in international politics—that did not ideologically compromise with capitalism. In 1971, the pro-Soviet hard-liners managed to gain a majority in the Socialist Student League (Sosialistinen Opiskelijaliitto, or SOL), the national organization of the Marxist–Leninist university students.[14]

The strategic alliance of students and communists was ideologically legitimized, firstly, by Leninist party theory (which held that only the Communist Party could combine the Marxist theories with revolutionary practice),[15] and, secondly, since neither the Finnish nor the Soviet communists considered the Finnish situation ripe for an immediate revolution, their intention was to create an 'anti-monopolist front' to prepare for socialism within the capitalist system. Essentially, this meant an emphasis on the control of the state organs. Here the Leftist radicals continued the long, state-centric tradition of Finnish—and Nordic—political culture, preferring the state to civil society as the main arena of political action.[16]

In all, the Marxist–Leninist student radicals hoped to bring the class struggle to the universities by combining socialist studies now with professional practices in future. This strategy was supposed to solve 'the dilemma of socialists with a master's degree'—the isolation of the Leftist educated class from

---

12 Jalava 2012, 86f.

13 Thomas Ekman Jørgensen, 'Scandinavia', in Klimke & Scharloth 2008, 239–252 at 241–244.

14 Matti Hyvärinen, *Viimeiset taistot* (Tampere: Vastapaino, 1994), 26–27; Risto Alapuro, 'Taistolaisten puolustus', in Johan Bäckman (ed.), *Entäs kun tulee se yhdestoista? Suomettumisen uusi historia* (Helsinki: WSOY, 2001), 339–343 at 339 and 342.

15 See, for example, Kari Toikka, 'Puolueesta', *Soihtu*, 2 (1971), 25–46.

16 Jalava 2012, 90–91. For the state-centric tradition in Finland and the other Nordic countries, see also Norbert Götz, 'Century of Corporatism or Century of Civil Society? The Northern European Experience', in Norbert Götz & Jörg Hackmann (eds.), *Civil Society in the Baltic Sea Region* (Aldershot, Hampshire: Ashgate, 2003), 37–48 at 39ff.

the politics of the working class—by showing how to unite theoretical thinking with party-political commitment and daily work. As a result, when the Western far-Leftists went to the barricades or armed themselves with machine guns, their Finnish comrades exhausted themselves with endless organizational duties. While the original objective had been the combination of university studies with party-political action, in many cases the activists ended up as full-time party *apparatchiks*.[17]

The University under Stamocap Attack

In the Finnish context, the 'proletarian turn' of the Leftist student movement has been explained in a variety of ways, ranging from a counter-reaction to inter-war right-wing nationalism ('the White Lie'), and increased competition for career opportunities, to symbolic patricide, and even a collective mental disturbance.[18] While all these explanations might be correct on the individual level, there is an evident lack of historical analyses of the theoretical basis of the Marxist–Leninist student movement in the 1970s. However, while it is not my intention here to deny the deeply problematic nature of the movement—think only of its uncritical, if not downright unethical, attitude towards the totalitarian regimes such as the Soviet Union and the German Democratic Republic—if we are to understand its broad support, some attention must be paid to its ideology (including the idea of the university) without over-psychologizing it or dismissing it as a mere intellectual fad. By placing the everyday concerns of the students in the broad perspective of the capitalist system, SOL's leaders offered not only a convincing enough explanation for the current situation, but also the means to alter it.

Broadly speaking, the critical Leftist analysis of the university owed much to the theory of state monopoly capitalism (stamocap),[19] which assumed that the

17 Aarnio et al. 1970, 214–221; Hyvärinen 1994, 224–228, 288–293, 334–344; Kimmo Lind, *Radikalismista aikalaisanalyysiin: Tutkijaliiton kehitys 1970-luvulta 1990-luvulle* (Jyväskylä: Kampus Kustannus, 1998), 31–39.

18 See, for example, Jukka Relander, 'Taistolaisuuden psykohistoriaa', in Sari Näre (ed.), *Tunteiden sosiologiaa II* (Helsinki: SKS, 1999), 190–226 at 190–218; Alapuro 2001, 339–342; Anna Kontula, 'Taistolaisuus puberteettikapinasta takinkääntöön', *Historiallinen Aikakauskirja*, 2 (2004), 233–243 at 234.

19 In addition to Soviet theoreticians, the Finnish stamocap theories were inspired by the ideas of the Belgian-based Marxist economist Ernest Mandel, the stamocap project of the French Communist Party, published in two-volume *Traite marxiste d'economie politique: Le capitalisme monopoliste d'etet* (1971), and the East German *Imperialismus der BRD*

capitalist system had entered a qualitatively new stage in its history after the Great Depression of the 1930s and the Second World War. While the previous stage had also been characterized by monopolies—that is, the concentration and centralization of capital in the hands of the large transnational companies intent on controlling the markets—late capitalist Western society was *state* monopolist, in the sense that the interventionist welfare state strived to balance the effects of market forces; pre-empt serious social crises; and ensure those forms of production considered vital to the national economy, but too expensive or unprofitable for private entrepreneurs. As a result, the state had become the most important instrument of 'bourgeois class domination'. In the case of Finland, national income policy agreements were considered the latest example of this development. The profits of monopoly capitalism were private, the losses socialized, as the stamocap theorists summed up the prevailing situation.[20]

According to Leftist higher-education theorists—Yrjö Engeström, Antti Kasvio, Kari Toikka—the state-monopolist stage of capitalism had a decisive effect on the education system in general and higher education in particular. In any society, the basic tasks of education were to produce a labour force, to select individuals for certain professions and positions of power, and to socialize them into prevailing norms and values. However, in the late capitalist era, which was increasingly based on automation and high technology, the universities were acting not only as the producers of a skilled labour force and the instigators of bourgeois indoctrination, but also as 'knowledge factories', which produced new, economically valuable knowledge that could be bought and sold in the market. Thus, the monopolies had become interested in the universities in a new way, hoping to benefit from the results of scientific research. Nevertheless, while the costs were high and did not usually bring an immediate profit, in the stamocap system it was the state that increasingly invested in the branches of science that were useful to economic life. Although this meant a massive transfer of payments from tax-paying wage-earners to monopoly capitalists, the politicians and education reformers mostly denied the political character of these issues and presented them as mere technical decisions based on impartial expertise.[21]

(1972); see Heikki Mäki-Kulmala, 'Vamokap', *Työväentutkimus* (2006), available at <http://www.tyovaenperinne.fi/tyovaentutkimus/tt2006/a_vamokap.htm>, accessed 13 December 2011.

20 Timo Linsiö, 'Työväenluokka ja opiskelijat', *Soihtu*, 1 (1969), 4–6 at 5f; Yrjö Engeström, *Koulutus luokkayhteiskunnassa* (Jyväskylä: Gummerus, 1970), 41 & 85–88; Antti Kasvio, 'Valtiomonopolistinen kapitalismi Suomessa', *Soihtu*, 4 (1972), 3–63 at 10–15 and 31f.

21 Aarnio et al. 1970, 72–99; Engeström 1970, 79–80, 114–119 and 242–253.

In the eyes of Leftist student activists, the fusion of the monopolies, the state, and the universities was most profound in the US, where, as Thorstein Veblen had showed in his *Higher Learning in America* (1918), the capitalists had dominated the universities ever since their founding. In Western Europe, the recent degree reforms of West Germany (*Neuordnung des Studiums*), France (the Fouchet Plan), and Sweden (UKAS) showed similar tendencies, albeit seasoned with the restrictions brought by the feudal, professoriat-centred structures of the traditional European universities.[22] In the case of Finland, due to its late industrialization and the strong 'archaic humanistic tendency' of the professoriat, the qualitative transformation of the universities from the pre-industrial to the industrial mode of knowledge production had barely started. Nevertheless, even in Finland the state had adopted the new planning ideology that highlighted the needs of the economic life. Hence, it had challenged the hegemony of the traditional university elite by supporting pro-reform technocrats who perceived higher education as little more than an investment in the promotion of economic growth. In the late 1960s, this ideology had resulted in efforts to differentiate vocationally oriented basic degrees from advanced scholarly education, to favour the natural sciences and technology at the expense of the humanities, and to shorten study times, since 'to the capitalist, every year which the student spends away from production, away from producing surplus value for the capitalist, is a loss'.[23]

At the same time, the Marxist–Leninist students distanced themselves from the Humboldtian idea of the university. In principle, they sympathized with its demands for a critical, self-assessing attitude, the unity of teaching and research, and the freedom of teaching and studying. Nevertheless, since some historically minded Leftist activists—above all, the historian Matti Viikari—were not only interested in Marxism, but the German history of ideas in general, they insisted that the Humboldtian tradition had to be critically re-evaluated in its proper historical context. In their view, as a state institution, the Humboldtian university was born as a part of the mechanism intended to defend the leading position of the feudal upper classes against the ascending bourgeoisie. Thus, knowledge that was produced at the nineteenth-century German-type universities was deliberately impractical for the immediate needs of productive forces, whereas it was politically and ideologically expedient in the restoration of the feudal state structures after the Napoleonic Wars. This restoration included the creation of national culture and the education of étatiste civil servants, who perceived themselves as a non-partisan force

22 Engeström 1970, 247–253.

23 Aarnio et al. 1970, 95–102 and 210f.

above the class conflicts of civil society. In return, the universities were allowed an inner autonomy which, being based on an unhistorical, idealistic theory of the state, came to mean the autonomy of professors within the feudal–hierarchical structures of the university. 'False consciousness' was formed, since the very idea of academic autonomy was identified with bourgeois apoliticism and the privileged position of the professoriat. Hence, even in the latter part of the twentieth century, the professoriat stubbornly opposed students' participation in the university administration, failing to notice that the stamocap onslaught was eroding the significance of traditional academic decision-making arenas.[24]

In the beginning of the 1970s, the higher-education system was considered to be faced with an inevitable transformation. On the one hand, the economy was increasingly based on high technology and science as a productive force, which required massive investments in higher education. Since the monopolies were not willing to invest their own capital income, their solution was to lobby for reforms aiming at greater efficiency and rationalization—more results at the same price. This tendency had increased the specialization, technicalization, vocationalization, and instrumentalization of academic education and research, bringing the position of university graduates closer to that of non-academic skilled workers, and the professoriat closer to administrative and financial management. On the other hand, the future development of production and working life posed new demands for all workers: the capacity for broad conceptual thinking, flexibility, innovativeness, language skills, and the ability to cooperate with other people were essential, whether the political system was capitalist or socialist. Moreover, these qualities would be essential for the emergence of revolutionary subjectivity and 'democratic professional practices' ('demokraattinen ammattikäytäntö'), which would be important when the Leftist students graduated and set out to change the world through their daily work. Since it was obvious that neither the Humboldtian university model, which was hopelessly detached from vocational and productive realities, nor the 'pressuring methods' of the pro-reform technocrats, who were in the direct service of the stamocap regime, could effect the necessary transformation, SOL demanded reforms that avoided both the Scylla of 'archaic humanism' and the Charybdis of 'technocracy'.[25]

In the most utopian visions, inspired by the experiments in the 'social laboratories' of Cuba and China, developments showed a promising tendency towards integrating universities into Finland's 'progressive' social life as a

---

24 Ibid. 107–136; see also Engeström 1970, 46f and 61.

25 Aarnio et al. 1970, 130–145, 210–216; Engeström 1970, 95–108, 121f. and 169f.

whole, thanks to which the isolation and alienation of students and academic staff were overcome. A case in point was the Faculty of Social Sciences at the University of Havana. In 1970, the whole faculty had moved for seven months to the Camilo Cienfuegos Sugar Mill, working by day and studying in the evening. As Fidel Castro had defined the overall objective in 1968, '*la Universidad se universaliza*', the university makes itself universally available, factories, mills, farms, forestry plantations, and other workplaces would be made into universities offering life-long higher education to all citizens.[26]

The Humboldtian Revival

Since the idea of taking over the state machinery was a strategy favoured by the Finnish Leftist student activists, they participated in the early 1970s in university reforms that seemed to offer an opportunity to change the system. To some extent, the centre–left government was inclined to heed their demands. On the one hand, the state authorities could play on the student movement's antipathy towards 'the professorial oligarchy', thereby exploiting the existing conflict at the universities for their own purposes.[27] On the other hand, the politicians and civil servants had been scared by the violent student revolts in Europe, and now sought contact with the young in order to prevent the escalation of protests in Finland, relying on a strategy of integration to tame radical currents at the university.[28]

At first, an 'unholy' alliance of centre–Leftist politicians, civil servants, and student radicals tried to push through a radical, administrative reform of the universities, based on the 'one man, one vote' principle—in other words, on a demand for students' and academic staff's equal voting rights in university elections.[29] Later, the student movement supported the university degree reform, which seemed to be moving in a 'progressive' direction under the leadership of Yrjö-Paavo Häyrynen, an educational psychologist and a representative of the Social Democratic Party's left wing. Although the degree reform

26 Mikko Pyhälä, 'Korkeakouluopetuksen tavoitteet Kuubassa', in Yrjö-Paavo Häyrynen, Jussi Jyrkinen & Paavo Kosonen (eds.), *Tutkinnon uudistus* (Helsinki: Valtion painatuskeskus, 1971), 132–137 at 132–136.

27 See also Elzinga 1993, 215ff.

28 Esa Sundbäck, *Suomen Ylioppilaskuntien Liitto ja suomalaisen opiskelijaliikkeen muutoksen vuodet 1968–1990* (Helsinki: SYL, 1991), 36f; Veli-Matti Autio, *Opetusministeriön historia*, vi. (Helsinki: Opetusministeriö, 1993), 20–23 & 481–484.

29 For a detailed discussion in English, see Matti Klinge, *A European University: The University of Helsinki 1640–2010* (Helsinki: University of Helsinki, 2010), 774–785.

committee did not adopt a Marxist framework, it envisaged a polytechnic degree model that could be interpreted as a means of realizing the idea of 'democratic professional practices'. The degree model was supposed to combine, on the one hand, scholarly education with vocational and social competences, and, on the other, theoretical approaches with practical problem-solving abilities. Moreover, Häyrynen's committee openly criticized the technocratic reform plans of the 1960s, and thus could be considered a counterweight to the stamocap element at the universities.[30]

However, the times were changing. Although signs of crisis in the capitalist world had been evident for some time, it was the 1973 oil crisis that exacerbated the economic difficulties, simultaneously causing high inflation, mass unemployment, and decelerating growth. Finland was no exception, and despite the buffering effect of its bilateral trade with the Soviet Union, by 1974–1975 it was facing an economic downturn. This meant budget cuts for the higher-education sector, including degree reform. For example, because of cutbacks, the pedagogical dimension of the reform was largely missed, whereas the immediate relevance of higher education to the labour market gained in urgency. University staff were particularly worried about the academic quality of the new polytechnic degree programmes, claiming that these 'narrowly occupational degree tunnels' had no disciplinary cohesion, and so endangered the existence of the university as a scholarly community.[31]

The growing discontent with the state-led higher-education policy in general and degree reform in particular took on more organized forms at the Faculty of Social Sciences at the University of Jyväskylä, which had conducted a pilot experiment in new degrees. In 1976–1977, the opposition gathered under the banner of the Humboldtian *Bildungsuniversität*. As a part of a more general reaction against Anglo-American, positivist, behaviourist, and empiricist currents in the social sciences and the humanities, the defenders of the *Bildungsuniversität*—students and academic staff at the departments of Philosophy and Sociology prominent among them—returned to the Continental European traditions, which led them to rediscover the neo-Humanist, hermeneutic, and idealistic university ideas of the German philosophers, Wilhelm von Humboldt, Immanuel Kant, Friedrich Schleiermacher, and Johann Gottlieb Fichte.[32] They were equally inspired by 'the Humboldt of

30 See Jalava 2012, 104–109, 126f.

31 Ibid. 111ff, 141f.

32 For a brief historical introduction to the Humboldtian idea of the university, see, for example, Esa Sironen, 'Humanistisesta ja tieteellisestä yliopistosta', *Sosiologia*, 5 (1977a), 211–214.

Finland', the philosopher Johan Vilhelm Snellman. Of special importance was Snellman's lecture *Om det akademiska studium* ('On academic study'), published in 1840. While Snellman, much like the world-famous German philosophers, had spoken up against Enlightenment utilitarianism and its instrumental view of higher education, the opponents of the degree reform felt a kinship with him and linked themselves with the long Finnish–German tradition of *Bildung*.[33]

For the Leftist advocates of the *Bildungsuniversität*, the seminal idea of the Humboldtian university was the principle of *Bildung durch Wissenschaft*, which can be roughly translated as 'self-formation through erudition'. According to this ideal, the goal of university studies was to initiate students into scholarly thinking. As Immanuel Kant had suggested in his essay 'Beantwortung der Frage: Was ist Aufklärung?' ('Answering the Question: What is Enlightenment?') (1784), the students had to learn to use their own understanding and reason publicly in all matters. This *Bildung* process required academic studies and research to be freed from demands for immediate social relevance and narrow occupational specialization, and instead be based on an interest in scientific knowledge and the quest for truth. In the long term, this kind of higher education was also the most advantageous to society and the state, as it educated students to become critical, civilized, and ethical individuals, who were able to transcend prevailing conditions.[34]

Initially, SOL's members were divided on the issue. According to the conciliatory view, the original targets of the degree reform had been accurate, but its implementation was a failure.[35] However, many members of the Leftist student movement were open to more profound self-criticism. As they put it, it had been naïve from the outset to think that truly progressive planning of the university degrees was possible within state monopoly capitalism. When SOL had accepted the validity of the occupational qualifications framework and production structures as the basis of reform, it had actually subscribed to the demand for a manpower analysis of the kind typical of the OECD's capitalist higher-education policy. Moreover, while insisting that universities should be at the service of the people, the democratic student movement had adopted an instrumental view of higher education, which in Finnish capitalist conditions in practice served stamocap interests.

---

33 Ibid.

34 Ibid.; Matti Juntunen, 'Yliopistolaitoksen synty ja tilanne tänä päivänä', *Kansan Uutiset*, 22 October 1978.

35 Antti Kasvio, 'Teesejä tutkinnonuudistuskeskusteluun', *Tiede & Edistys*, 4 (1977), 53–57 at 56f.

By 1978, the majority of SOL's members were already demanding that degree reform should be immediately terminated and the autonomous status of the *Bildungsuniversität* acknowledged as the basis of all future reforms. The new guidelines should follow the internal logic of academic study and research based on intellectual growth and *Bildung*, not the external logic of party politics and the labour market.[36] For those communist hard-liners who had difficulty swallowing such early nineteenth-century German idealism, encouragement was offered by their East German comrades, such the Rector of the Martin Luther University of Halle-Wittenberg, Dieter Bergner. He assured a Finnish audience in the SOL magazine *Soihtu* ('The Torch') that the universities of the German Democratic Republic were actually the true descendants of the most progressive features of Humboldt's ideal university.[37] After this U-turn, the Marxist–Leninists somewhat embarrassingly found themselves on the same side as the traditionally minded academic professoriat. After a decade of constant quarrels, almost the whole academic community was once again gathered on the same side of the barricades to defend the core academic values associated with Humboldtian ideals.

The Humboldt Myth

In current debates over the university system, the essence of Wilhelm von Humboldt's ideals and their relevance in the founding of the University of Berlin in 1810 have been contested issues. As the sociologist Georg Krücken and others have remarked, while it is obvious that Humboldt's conception of the university can hardly be taken as an accurate description of what universities actually are in an era of mass education, it is equally debatable whether German universities—or the Nordic ones modelled on them—have in any phase of their history been truly Humboldtian institutions. Hence, Krücken interprets the tenets of Humboldt's university concept—the unity of teaching and research; social disembeddedness and autonomy; a non-utilitarian approach to higher education as opposed to vocational training—as an organizational myth of the modern research university.[38]

36 See, for example, Juha Manninen, 'Teesejä sivistysyliopistosta', *Joukkovoima*, 5 (1978), 3f; Juha Sihvonen, 'Tutkinnonuudistus ja demokraattinen opiskelijaliike', *Soihtu*, 6 (1978), 6–24 at 14–23.

37 Dieter Bergner, 'Sivistysyliopisto tänään', *Soihtu*, 1 (1979), 5–22.

38 Georg Krücken, 'Learning the "New, New Thing": On the Role of Path Dependency in University Structures', *Higher Education*, 46 (2003), 315–339 at 324–327.

To be sure, faced with continuous academic specialization, one can hardly cherish the kind of holistic thinking in broad cultural categories that inspired Humboldt and his like-minded colleagues two hundred years ago. Moreover, when countries such as Finland offer free university education to tens of thousands of young people every year, it would be ridiculous to suggest that this academic education should concentrate on 'pure' science and 'impractical' knowledge, without the slightest idea of what all these people are going to do once they have graduated. Even if one would like to step back from the instrumentalism that subordinates higher education to immediate labour market needs, one certainly does not have to be a fervent supporter of the theory of state monopoly capitalism to agree with past higher-education theorists of the Left that a university cannot be based on reasoning that is totally detached from the vocational and productive realities of late modern society.

However, as noted by Krücken, and indeed some of the Finnish Leftist advocates of the *Bildungsuniversität* in the 1970s, the mythic, unrealistic character of Humboldt's ideals does not necessarily nullify their ongoing relevance.[39] An alternative reading is offered by the Frankfurt School, and especially by Herbert Marcuse, who in his attempt to contribute to Marxist aesthetics paid special attention to the Idealistic and Romantic literature of the eighteenth and nineteenth centuries. According to Marcuse, Marxist criticism often scorned the 'inwardness' and 'autonomy' of idealistic subjectivity as 'bourgeois' notions—qualities that were also central to the Humboldtian ideal of the university—and instead favoured realism in various senses as the intellectual framework that corresponded most closely to social relationships, in Marcuse's eyes easily taking on the colouring of vulgar materialism. For him, the radical and political potential of idealistic literature lay precisely in its concept of art as illusion (*Schein*), something that was 'unreal' in the ordinary sense of this word, but, first and foremost, qualitatively 'other' than the established reality. As such, art could protest and transcend given social relations as well as subvert ordinary everyday experiences, thus opening a dimension in which human beings and things no longer stood under the law of the established reality principle.[40]

In his discussion of the meaning of the Humboldtian *Bildungsuniversität* in 1977, the Finnish student activist Esa Sironen explicitly deployed this Marcusean idea of two-dimensional thought, in which 'fictional', 'unrealistic' idealistic thinking formed a critical corrective to the actual conditions of the

39 See Esa Sironen, 'Humboldtilainen yliopistonäkemys', *Keskisuomalainen*, 3 March 1977(b); Krücken 2003, 324–327.

40 See, for example, Herbert Marcuse, *The Aesthetic Dimension: Toward a Critique of Marxist Aesthetics* (Boston: Beacon Press, 1978).

university and higher education.[41] From this perspective, we may say that the 'Humboldt myth' could act—and still acts—as another dimension of a given reality, and its continuous importance stems precisely from the fact that it is irreconcilably antagonistic to the everyday order of university business, thus emphasizing the tension between the actual and the possible. As such, it helps to evaluate both the current state and future trends in higher education, and provides academic staff and students with meaning in the shape of a shared identity and subjectivity, which shields them from external influences and political pressures that violate their sense of what universities ought to be.[42]

41 Sironen 1977b.

42 See also Herbert Marcuse, *One-Dimensional Man* (London: Routledge, 2007) (first pub. 1964), 60–66 and 136f; Krücken 2003, 325ff.

PART 3

Contemporary Contentions

CHAPTER 8

'Humboldt' Humbug! Contemporary Mobilizations of 'Humboldt' as a Discourse to Support the Corporatization and Marketization of Universities and Disparage Alternatives

Susan Wright

In debates about the future of universities, 'Humboldt' is a key-word, a site of contestation, which is deployed by some to disparage the past and by others to decry a marketized future. Each side links 'Humboldt' to 'freedom', but in different ways. 'Pro-marketeers', to use the Danish catchphrase, are 'setting universities free' from the government interference of the Humboldtian past, to become 'free agents' in a knowledge economy. Pro-marketeers project the knowledge economy as an inevitable and fast-approaching future, and envisage universities as negotiating the demands that come at them from all sides for the precious raw materials of that economy—research and graduates. They caricature the discarded past as 'Humboldt', referring to the universities' collegial rule and critical knowledge pursued for its own sake. But these characteristics, protected by a range of freedoms—freedom from external interests, freedom to teach, freedom to learn, freedom of expression, research freedom, academic freedom—are what academic activists hold dear as the means to sustain the core intellectual and social values of a university. In recent books, geared to generate public debate in Denmark, academics have tried to rework the characteristics attributed to 'Humboldt', but why have they not succeeded in reasserting these characteristics in their own right as an alternative image of the future university?

In the struggles over university reform in Denmark from 2003 to the present, this contest over 'Humboldt' has been far from evenly matched. I will explore how it has played out in policy texts, notably those from the OECD, on which the Danish Government has drawn heavily; in international agreements over academic freedom, to which academic activists refer; and in the polemic genre of books and newspaper articles that activists have authored in order to generate public discussion. These issues have been discussed in interviews with policy-makers and academics, which were conducted as part of a research project on the university

 | DOI 10.1163/9789004271944_010

reforms.[1] Finally the chapter draws on participant observations at an event that acted out this contest over 'Humboldt', 'freedom', and divergent visions of the future university. Inspired by Stuart Hall's analysis of a political strategy he calls the 'double shuffle', I will argue that those advancing a vision of the university modelled on the corporate world and acting in global markets can only do so if they simultaneously deploy an image of what they call the traditional or 'Humboldt' university. This is a ploy to reassure academics and students that academic values will survive and to channel their energies into the reforms. With each new step, as the market discourse becomes more established and assured, the supporting discourse—the 'Humboldtian' vision—becomes increasingly caricatured and bowdlerized. The continual dance between the market and the 'Humboldt' visions of the university pushes forward the agenda for university reform.

Why have recent attempts failed to de-couple 'Humboldt' from this subordinate position in the dance? Appealing to international declarations to which Danish ministers and university leaders are signatories, Danish academic organizations and activists have succeeded in getting some legal modifications, but have failed to unsettle the dominant discourse, let alone reassert 'Humboldtian' values as a valid alternative. I would argue that those promoting university corporatization and marketization are giving a liberal twist to the meanings of 'freedom' and 'autonomy', but without spelling out how these key-words code their implicit world-view. The academic opposition are not reading this code: they assume that these key-words still stand for the values they hold dear. The two parties talk past each other, and the academic opposition does not expose or contest the growing hegemony of the liberal market world-view. As a result, proponents of corporatization and marketization can disparage the academic opposition for peddling a high-minded vision of a golden age that never existed, or was never golden, and for making grandiose statements about what should happen in future, which stand no chance of being put into practice—mere humbug!

The Double Shuffle

Beginning in the late 1990s, Denmark has been an avid supporter of the OECD's argument that its members should prepare for the inevitable future of a global

1 The project, 'New Management, New Identities? Danish University Reform in an International Perspective' was funded by the Danish Research Council (2004–2009).

knowledge economy. The ability of OECD members to maintain their competitive edge as the richest countries in the world depends on the speed with which they generate new knowledge and technology, and translate them into innovative products with high added value. This means that governments have to ensure their education systems, and especially their universities, produce the knowledge, technologies, and graduates on which their country's prosperity depends.[2] According to this argument, the rest of manufacturing can be relocated to cheaper sites in the global south.

The OECD has produced reports, checklists for action, and guidance on best practice so governments can reform universities to meet the needs of the knowledge economy.[3] Through a sequence of scenarios about the future university,[4] it is possible to trace a series of moves from an initial 'market' model of the university (akin to new public management, with the state playing a role in moderating the market and assuring quality), via the 'entrepreneurial university' and commercial 'edutainment', to a future scenario called HE Inc., where universities shed their traditional academic values, government funding, and social agendas (for example, commitments to equity). In HE Inc., the state is absent, each university seeks out its own niche in world competition and rankings, and there are no brakes on free trade.

Each of these progressively more market-oriented models is accompanied by one or more moderate versions. In the first of these reports, the 'instrumental or market' model is paired with one called variously 'Humboldt', 'critical intellectual', or 'the independent and unconstrained advancement of knowledge'.[5] Close analysis of this report shows that no attempt is made to reconcile the 'market' and 'Humboldt' models.[6] The 'market' model is always described

2 Danish Government, *Progress, Innovation and Cohesion: Strategy for Denmark in the Global Economy* (The Prime Minister's Office: Copenhagen 2006).

3 OECD (Organisation for Economic Cooperation and Development), *Redefining Tertiary Education* (Paris: Organisation for Economic Cooperation and Development 1998); Miriam Henry, Bob Lingard, Fazal Rizvi & Sandra Taylor, *The OECD, Globalisation, and Education Policy* (Oxford: Pergamon 2001).

4 OECD 1998; OECD (Organisation for Economic Cooperation and Development), *OECD/CERI Experts Meeting on 'University Futures and New Technologies'* (Washington: World Bank 2005); (Organisation for Economic Cooperation and Development), 'Four future scenarios for Higher Education', in *Higher Education: Quality, Equity and Efficiency. Meeting of OECD Ministers of Education* (Athens: Greece 2006).

5 OECD 1998, 10.

6 Susan Wright & Jakob Williams Ørberg, 'The Double Shuffle of University Reform – The OECD/Denmark Policy Interface', in Atle Nyhagen and Tor Halvorsen (eds) *Academic Identities – Academic Challenges? American and European Experience of the Transformation of Higher Education and Research* (Newcastle upon Tyne: Cambridge Scholar Press 2012).

first. The 'Humboldt' model always starts with negative contrasts to the 'instrumental or market' model, promoting the latter to the status of a norm:

> The starting point [for the Humboldt model] is neither economic or social utility, nor the student as consumer, nor the institution as service provider; it is, rather, the academic community's mission of knowledge creating and disseminating.[7]

The 'market' and the 'Humboldt' models are associated, respectively, with managerial and collegial governance. After claiming, airily, that these two forms of management can be reconciled, but without saying how, the report concentrates on implementing the 'market' model.[8] Scorned, 'Humboldt' becomes an increasingly shadowy presence, losing detailed features, but never allowed to disappear completely: it hovers as a faint glimmer of hope in the background. In the second OECD document, where six scenarios are offered, what was previously called 'Humboldt' has become 'tradition' and a caricature of 'universities as they are today': catering for an elite and regulated by government with 'little opportunity for profit-generating activities'.[9] This account emphasizes shortcomings, omits strengths, and invites dismissal. In the third document, four scenarios are offered.[10] Now 'Humboldt' and 'tradition' have gone completely and the most moderate scenario is called 'New Public Management', where universities use their autonomy to reconcile the demands of government and the markets. In sum, over this sequence of documents, the 'market' model steps progressively in the direction of global free trade, with market imperatives steering universities and little or no role for government, let alone collegial rule. But the market model is not alone. As Stuart Hall makes clear, in the double shuffle bold political moves towards a market state or a free market can only be made by a matching step that simultaneously reassures and carries along those who seek signs of hope that traditional values will not be extinguished.[11] Thus the most extreme market model is always accompanied by a more moderate scenario, which also sows confusion, obscures the long-term objective, and impedes the emergence of a coherent opposition.

7 OECD 1998, 45.

8 Ibid. 83.

9 S. Vincent-Lancrin, 'Building Futures Scenarios for Universities and Higher Education: An International Approach', *Policy Futures in Education,* 2 (2004), 259.

10 OECD 2006.

11 Stuart Hall, 'New Labour's Double-shuffle', *Soundings: a Journal of Politics and Culture,* 24 (2003).

The moderate scenario is not only subordinate to and dependent upon the dominant one, but to keep in step it must progressively transform by adopting more and more features of the advancing market model. As the double shuffle progressed through the OECD documents, the free market model finally split from new public management. This is a surprising separation, as new public management is usually associated with the concepts and technologies by which neo-liberal logic first impinged on institutions and professional practices. New public management now took on the mantle of tradition, to which it was previously opposed. It adopted the reassuringly subordinate position, appealing to the vestiges of university values. What started as 'Humboldt' becomes briefly 'Tradition', equated with 'universities as they are now'—a temporal fixity which disqualifies them as serious scenarios for the future—and finally the subordinate, reassuring position is taken by new public management. By this time, 'Humboldt' and 'tradition' are no longer even ridiculed on the sidelines: they have disappeared as an unviable alternative, no longer keeping up with the dance or even worth caring about—'Bah, humbug!'

'Freedom' in the Market Discourse

This double mode of address, for pro-marketeers and 'traditional' academics alike, hinges on the use of key-words, notably autonomy and freedom, in the sense that they are 'keys' to a contest over how universities should be envisaged and organized. University autonomy and freedom have accumulated a range of associations, some connected to the market model and others to 'Humboldt'. In Walter Bryce Gallie's and Raymond Williams's terms, key-words are continually contested, and they never have a complete or final meaning.[12] The rival meanings of a single key-word are at their clearest in the clusters of other words with which they are associated. In the present case, two 'semantic clusters' around 'freedom' pull in opposite directions—towards the market discourse and 'Humboldt' alike—but as long as the differences between them are not made clear, the market discourse avoids being exposed and effectively contested: on the contrary, as will be shown, it manages to mobilize the positive charge associated with Humboldtian 'freedom' and transfer it to its very different, even contrary idea. And so the double shuffle progresses.

When the Danish minister for research announced that, with the 2003 Universities Act, he was 'setting universities free', this did not signal that

12 W.B. Gallie, 'Essentially Contested Concepts', *Proceedings of the Aristotelian Society* 56 (1956); Raymond Williams, *Keywords* (London: Fontana Press 1976).

universities would be more independent and less constrained by government. On the contrary, it heralded what Nikolas Rose calls 'government through freedom'.[13] This liberal, political rationality envisages the space to be governed as divested of direct state control and constituted of free subjects, whether they are individuals or institutions. The art of government is to deploy technologies and strategies that shape the sense of a moral, responsible, accountable subject, who then uses its own creativity, capacities, and sense of freedom to produce the ends of government. As Rose puts it, subjects are obliged to be free in particular ways.[14]

The first step in setting universities free was to change their legal status to 'self-owning institutions', which the Danish ministry translated (inaccurately) into English as 'institutional autonomy'. In some of our interviews, academics assumed that this meant universities would no longer be state institutions, would be free of heavy state interference, and have a status akin to the UK's publicly funded independent corporations. In fact, theirs is a very ambiguous status: universities are still come under the state, but they now have separate capital (*særeje*) and their own audited accounts within the state's consolidated accounts. Their land and buildings are, in the main, still owned by the state, and the state largely controls their liquidity, yet they are responsible for their own solvency. The 'self-owning' university is 'set free' to chart its own course as a largely publicly funded agent or resource in the global knowledge economy. Government, industry, and other stakeholders in 'surrounding society' can all make demands on universities for the research and teaching that meets their needs. It is up to the 'self-owning' university to decide which demands to accept. When, soon after the Act was passed, the minister of education visited the Danish University of Education to argue that his political priorities should feature strongly in the institution's research agenda, *ForskerForum*, the magazine of the researchers' trades unions, expressed shock and outrage. But the minister was acting according to the new logic of the 'free' university: he was 'free' to make demands, and the university was supposedly 'free' to determine its own response.[15]

While 'Humboldt' is often associated with the state protecting its universities from political and economic demands from society at large, one

13 Nikolas Rose, 'Towards a Critical Sociology of Freedom', Inaugural Lecture Goldsmiths College, London, 5 May 1992.

14 Nikolas Rose, *Governing the Soul: The Shaping of the Private Self* (London: Routledge 1989).

15 Wright, Susan & Ørberg Jakob Williams 'Autonomy and Control: Danish University Reform in the Context of Modern Governance', *Learning and Teaching: International Journal of Higher Education in the Social Sciences*, 1 (2008).

interviewee (a former ministry official) explained that their previous position in the machinery of state afforded universities little protection and incurred considerable and detailed bureaucratic intervention. Nevertheless, the 2003 Universities Act made clear that universities no longer existed in a state-protected space, for it stated clearly that universities were now responsible for protecting their own research freedoms and scientific ethics. That particular paragraph does not specify who 'the university' is, but a later paragraph states that the new governing board, with a majority of appointed, external members, is the highest authority and must safeguard the university's interests.[16]

Whereas the liberal strand of the reforms aimed to govern the university through the exercise of its own 'free' agency, in a second, more dirigiste strand, 'self-owning' universities were part of a strengthening the centralized political steering of the whole public sector in Denmark. It has been argued that new forms of government emerge from assemblages of different rationalities and technologies that are rarely coherent or consistent.[17] The ministry of finance, which led public sector reform, explained in a report that Denmark was shifting from a bureaucratic state to what they called 'aim and frame steering' with the aim of gaining tighter political control of institutions' activities and making them respond more quickly to changes in policy.[18] Henceforth the ministers would only focus on the political aims of public services, and the budgetary and legal frameworks for their achievement. In what Pollitt et al. call a process of 'agentification', state institutions were made into 'free agencies', which were then contracted to deliver these services.[19] Ministers were to use contracts, performance indicators, and payment by results to ensure that agencies carried through the political agenda. Universities' 'self-owning' status enabled them to enter into contracts with the education minister (as well as other stakeholders). The universities' fulfilment of their contractual commitments to the minister would then be checked annually by the state auditor. Anna Yeatman (1997), in writing about the contractual state, argues that the concept of 'freedom' used here is not that of classic liberalism. In classic liberalism, a contract is made between formally equal parties, both of whom can exercise

16 Folketinget, *Act on Universities*, Act no 403 of 28 May 2003, §§ 2, 10, available at www.rektorkollegiet.dk.

17 Nikolas Rose, Pat O'Malley & Mariana Valverde, 'Governmentality', *Annual Review of Law and Social Science*, 2 (2006).

18 Wright & Ørberg 2008, 37ff.

19 Christopher Pollitt, Karen Bathgate, Janice Caulfield, Amanda Smullen & Colin Talbot, 'Agency Fever? Analysis of an International Policy Fashion', *Journal of Comparative Policy Analysis: Research and Practice*, 3 (2001).

the freedom to choose who to enter into a contract with, on what terms, and when to leave. By contrast, the 2003 Universities Act *obliges* Danish universities to enter into a contract with the minister; in practice the ministry sets the parameters for the contract and the university is only able to negotiate the details. And the university is not free to terminate the contract.[20]

Policymakers used 'freedom' in a third way when they said that universities (*qua* their leaders) were given 'freedom to manage'. The 2003 Universities Act established a system for appointing rectors, deans, and heads of deparment (where previously they and university senates, faculty boards, and departmental boards were all elected by university staff and students). In many universities, contracts between the leaders progressively outsourced the minister's agenda at all levels of the organization. Leaders' contracts expected them *inter alia* to increase external funding and knowledge transfer to industry (the government's catchphrase for the reform was 'From Idea to Invoice'); raise the throughput of students and their employability in industry; and increase publications to gain 'world-class' status in international rankings. The idea was that once universities had a rector who could act as an interlocutor with government and industry, and who had a hierarchy of leaders accountable only to him, these leaders would use their capacity for strategic management and their expanding scope of action to deliver this agenda. At which point, government could entrust universities with increased funds to drive Denmark's competitiveness in the global knowledge economy—indeed the restoration of political trust in the universities was a dominant theme when Parliament debated the 2003 Act.[21] Government entrusted leaders with freedom to manage their organizations as they wished, as long as their university performed as required.

In a fourth use of 'freedom', in language reminiscent of colonialism, government suggested that, as universities learned how to run themselves as 'free' institutions, proved themselves capable of strategic leadership, and prioritized their activities and resources to meet the country's needs in the global knowledge economy, then they might incrementally win further 'degrees of freedom' from government control. To Yeatman, the shift from classic liberalism to neo-liberalism involves changing the status of those who were previously

---

20 Susan Wright & Jakob Williams Ørberg, 'Prometheus (on the) Rebound? Freedom and the Danish Steering System', in Jeroen Huisman (ed.) *International Perspectives on the Governance of Higher Education* (London: Routledge 2009).

21 Jakob Williams Ørberg, 'Trust in Universities – Parliamentary Debates on the 2003 University Law', *Working Papers on University Reform, No. 2* (Copenhagen: Danish School of Education, University of Århus 2006).

dependants of patrimonial governance, be it the family's or the state's.[22] These sometime dependants have to learn to think of themselves and manage themselves as autonomous units of social action if they are to acquire the capacity to act as freely contracting beings.[23] This was very much the language of the Danish government. However, the 'degrees of freedom' were very small and came very slowly, as exemplified by the minister's statement on 22 April 2009 that he would give universities a greater degree of freedom by reducing the level of detail in the contracts to a few goals, and by giving leaders greater freedom to choose how to fulfil them.[24] Some chairs of the new governing boards, with experience of directing major Danish firms and institutions, grew impatient with this paternalistic approach and complained that government had retained control of too many areas of decision-making, and made so many detailed interventions that it impeded their ability to do their job.[25] To the government, the intensity of state intervention and control was not a contradiction, but a matter of degree: if universities were steered into acting responsibly, efficiently and strategically, eventually these newly liberated subjects would learn how to use their freedom within a framework of less visible systems of government control.

Academic Discourses about 'Freedom'

Academics refer to freedoms that have very different histories and meanings to the ideas of freedom in the government discourse. In their campaigns to protect their freedoms, academics have referred to international agreements, especially those to which the Danish authorities are signatory.[26] These agreements focus on concepts of academic freedom, collegiality, and institutional autonomy associated with 'Humboldt'. But Danish academics often do not frame the issues in quite these terms, and do not draw on the full range and strength of these concepts.

---

22 Anna Yeatman, 'Contract, Status and Personhood', in Glyn Davis, Barbara Sullivan & Anna Yeatman (eds.), *The New Contractualism?* (Melbourne: Macmillan Education Australia 1997).

23 Ibid. 46.

24 Kristina Villesen, 'Sander vil give universiteterne mere frihed' *Information* 22 April 2009.

25 'Giv universiteterne fri...', *ForskerForum*, 202 (2007), 1.

26 For example Peter Harder, 'Kvalitet på universitetet: refom eller tilintetgørelse?', in Helge Sander (ed.), *Fremtidens universiteter* (Copenhagen: Gyldendal 2009), 115; and Ingrid Stage, 'On Academic Freedom and Freedom of Expression', *Magisterbladet*, 11 (2008), 28f.

In 2008 one of the academic unions, Dansk Magisterforening, submitted a formal complaint to UNESCO against the Danish government for not complying with its international standards of academic freedom. UNESCO's 1997 Recommendation on the Status of Higher Education Teaching Personnel defines academic freedom as

> the right, without constriction by prescribed doctrine, to freedom of teaching and discussion, freedom in carrying out research and disseminating and publishing the results thereof, freedom to express freely their opinion about the institution or system in which they work, freedom from institutional censorship and freedom to participate in professional or representative academic bodies.[27]

The term 'academic freedom' is rarely used in Denmark. Instead, academics refer to *forskningsfrihed* ('research freedom'), which usually refers to freedom to choose a research topic and an appropriate method. Other aspects of academic freedom, such as the connection between research and teaching through the lecturers' control over the curriculum with the freedom to decide what to teach and how, and students' educational freedom (the German concepts of *Lehrfreiheit* and *Lernfreiheit*),[28] are absent from these formulations, as is the British provision for academics to choose who should join their community as staff and students.[29]

The definition of 'research freedom' was tightly constrained in the 2003 Universities Act, and, after a long campaign by academics, was slightly revised in 2011. A clause qualifies the general statement that the university and the individual academic have research freedom by saying that academics' research must fall 'within the bounds of the university's research strategy' and they 'are free to perform independent research when not performing work assigned by management'. However, these tasks must not 'require the entirety of their working hours over longer periods of time, which would in essence deprive

27 UNESCO's report from 1997, § 27, quoted in Stage 2008.

28 The Ministry and the Accreditation Council set parameters for university education and approve each curriculum. For *Lehrfreiheit* and *Lernfreiheit*, see H.S. Commager, 'The University and Freedom. "Lehrfreiheit" and "Lernfreiheit"', *Journal of Higher Education*, 7 (1963).

29 In 1998 an executive order gave university leaders the authority to decide on appointments from among qualified candidates; and universities set the size of their student intake, but do not decide which students to admit.

them of their freedom of research'.[30] While an improvement on the previously even more draconian wording, research freedom still remains a residual category that is only defined by what it is not.

Meanwhile, it is still not clear *who* the university is and how protection of the university's and the individual's research freedom is to be achieved. When, after the 2003 Universities Act, Copenhagen University's new governing board was devising the university's statutes, the *Praksisudvalget* (committee on academic practice) submitted wording for the clause on research freedom:

> Research freedom involves freedom to choose a research topic, to pose questions, to decide which materials and which methods to use to find an answer, and freedom to present hypotheses, results and lines of reasoning publicly. The university will ensure that strategic research frameworks that are laid down in the development contract for the university's activities do not unduly restrict the research freedom of the individual employee. The university shall ensure that research assignments are organised and allocated so that each individual scientific employee as far as possible can themselves choose research topics. Finally the university shall ensure scientific openness in research environments so that diversity and mutual criticism can be freely developed.[31]

The Governing Board refused to include this statement on the grounds it was too 'comprehensive' for their 'minimalist' statutes. Instead, they asked the rector to return this wording to the committee, for them to include it in their terms of reference and publicize it broadly among the university population. This committee does not have powers of enforcement.

The UNESCO Recommendation draws a direct connection between academic freedom and collegiality and democracy. After defining academic freedom (quoted above) it goes on to say that 'personnel can [only] effectively do justice to this principle if the environment in which they operate is conducive, which requires a democratic atmosphere'.[32] This point is expanded in a section on 'Self-governance and collegiality', where it is stated that academics 'should

30 Folketinget, 'Lov om ændring af universitetsloven, lov om teknologioverførsel m.v. ved offentlige forskningsinstitutioner og lov om almene boliger m.v.', http://www.ft.dk/RIpdf/samling/20101/lovforslag/l143/20101_l143_som_vedtaget.pdf, § 14, 6.

31 Copenhagen University, 'Minutes of Governing Board Meeting No. 25 held on 18 April 2007'.

32 UNESCO 1997, § 27 quoted in Stage 2008.

have the right and opportunity [...] to take part in the governing bodies', 'to elect a majority of representatives to academic bodies', and 'to criticise the functioning of institutions, including their own'.[33] It states that collegiality means shared responsibility and the participation of all in decision-making 'regarding the administration and determination of policies of higher education, curricula, research, extension work, the allocation of resources and other related activities'.[34] This concept of self-governance as participation in decision-making stands in stark contrast to the Danish government's expectation that individual actors, whether people or institutions, can be steered to exercise their own freedom and manage themselves to achieve the government's aims.

Academics have made numerous statements regretting the loss of 'democracy', meaning the election of leaders and boards at all levels of the university, and *ForskerForum* has campaigned on the issue, turning Karran's international survey of the features of university management and governance associated with academic freedom into a 'freedom index' chart, in which Denmark scored zero.[35] In 2008 a petition signed by 6502 people (40 per cent of all Danish university academic staff) was delivered to Parliament demanding a new law that would protect freedom of research, freedom of expression, and the inclusion of academics in decision-making processes.

Following criticism from an international evaluation, the 2003 Universities Act was revised in 2011 to give university governing boards responsibility for ensuring that there is 'co-decision making and involvement of employees and students in important decisions'.[36] Yet it is unclear how the university's *enstrenget ledelse* ('unified management'), with each leader being accountable and having a 'duty of loyalty' to the leader above them, can be made compatible with staff and students having a real say in decisions. Exponents of the dominant discourse dismiss calls for real collegial involvement as a throwback to the dysfunctional university democracy of the past, even though two leading academics have argued that the old system had weaknesses and they are not calling for its return.[37]

---

33 UNESCO 1997, § 31 quoted in Stage 2008.

34 UNESCO 1997, § 32 quoted in Stage 2008.

35 Terence Karran, 'Academic Freedom in Europe: A Preliminary Comparative Analysis', *Higher Education Policy*, 20 (2007); Terence Karran, 'Academic Freedom: In Justification of a Universal Ideal', *Studies in Higher Education*, 34 (2009); 'Frihedsindeks: Ufrie danske universiteter', *ForskerForum*, 2006.

36 Folketinget 2011, § 10, 6.

37 Sune Auken, *Hjernedød: Til forsvar for det borgerlige universitet* (Copenhagen: Informations Forlag 2010), 33–34; Harder 2009, 110.

With very few exceptions, academic arguments in favour of collegial rule say it is necessary for the exercise of research freedom, conceived of as an *individual right*. This misses a much wider point made in international statements that collegial rule is a prerequisite for academic freedom, which, as a *public responsibility*, is to benefit society. Signed in 1988 by rectors of European universities (including six of Denmark's eight universities) to mark the nine-hundredth anniversary of Europe's oldest university, the University of Bologna, the Magna Charta Universitatum describes the university as 'an autonomous institution at the heart of societies'. It continues, 'to meet the needs of the world around it, its research and teaching must be morally and intellectually independent of all political authority and economic power', and both government and universities must 'ensure respect for this fundamental requirement'.[38] The Council of Europe's Recommendation 1762 on 'Academic freedom and university autonomy', similarly ties academic freedom closely to institutional autonomy in the sense of protecting the university from the intrusion of economic and political interests.[39] Both argue for these conditions not as a right for individuals but on the grounds that they are crucial for universities to fulfil their role in society. The Magna Charta depicts this role in terms of sustaining the humanist traditions of democratic society. The Council of Europe criticizes demands for universities to respond to the short-term needs of the market, and argues that the role of universities is to contribute to the definition and solution of current social and economic problems, whilst analysing them in a distanced, long-term, and critical perspective:

> The social and cultural responsibility of universities means more than mere responsiveness to immediate demands of societies and the needs of the market, however important it may be to take these demands and needs seriously into account. It calls for a partnership in the definition of knowledge for society and implies that universities should continue to take a longer-term view and contribute to solving the fundamental issues of society as well as to finding remedies to immediate problems.[40]

Rarely do Danish academics argue in the same vein as these international documents that academic democracy and university autonomy are crucial for

---

38 "Magna Charta Universitatum", 2008, http://www.magna-charta.org/library/userfiles/file/mc_english.pdf.

39 Council of Europe, "Recommendation 1762", 2006, 'Academic Freedom and University Autonomy', http://assembly.coe.int/main.asp?Link=/documents/adoptedtext/ta06/erec1762.htm.

40 Council of Europe 2006, art. 8.

universities to fulfil their responsibilities to society. One exception is Lucy Smith who, in a festschrift for a previous rector of Aarhus University, argues that universities are granted autonomy in order for its members to exercise academic freedom, and that this 'critical function' of disinterested scholarship and the unfettered and honest debate of critical issues is not a right but a responsibility.[41] Members of the academic community are granted academic freedom by society not as a right but as a duty, as for example when making research findings known even if they are unpopular or contrary to the wishes or policies of government. She argues against a contemporary 'subtle shift toward caution'.[42] Fulfilling this responsibility towards society, she argues, is 'central to the idea of what a university is' and is more important than ever when it may be threatened 'in universities that are market-driven or cooperating too closely with industry'.[43]

In sum, these international recommendations focus on 'freedom', but in contrast to the semantic cluster identified here in the government's and OECD's discourse, these recommendations associate 'freedom' with a quite different range of words—research, teaching, dissemination, debate, critique, ethics, social responsibility, autonomy, democracy. The Danish academics' emphasis on the general constitutional right to individual 'freedom of speech', instead of focusing on the collective responsibility to protect the university's 'research freedom', and their substitutions of 'research freedom' for 'academic freedom' and of 'individual right' for 'social responsibility', render the semantic cluster in Danish academic discourse a weaker version of the international one set out above. Nor does it make such a coherent argument that these freedoms are needed to fulfil the university's responsibilities to society. It is rare for it to be argued that each individual needs to be defended in order to protect the research freedom of the university, and, concomitantly, that the university's research freedom needs to protected in order for its members to fulfil their duty to society of providing impartial research and participating in public debate. Both international and Danish academic discourses, however, associate 'freedom' with creating the conditions for a critical space for research, teaching, and public communication, based on a disinterested and long-term perspective, which is in stark contrast to the government's vision of freedom as steering others' 'free' agency to achieve its own short-term plans for how to compete in the global knowledge economy.

---

41 Lucy Smith, 'The Academic Values', in Carsten Bach-Nielsen (ed.), *Dannelse, Uddannelse, Universiteter* (Århus: Århus University Press 2001).

42 Ibid. 283.

43 Ibid. 277.

Head-to-Head in Debate Books

Both partners in the double shuffle prize the word 'freedom', but what are academics' views on what is happening to a keyword that they hold dear, as it is deployed to new purposes and to serve a new organizational and governmental logic? Do they see how their discourses of 'freedom' have been recruited to play a supporting role in the double shuffle, and any attempt to assert them as valuable in their own right is dismissed as 'Humboldtian'? Have they managed to decouple 'Humboldt' from this dance, and reassert their meanings of 'freedom'? Have they contested the government's shifts in the meaning of 'freedom', and the formation around it of a semantic cluster as well as a discourse that in OECD circles has become hegemonic? An important site for this contest over the concept of 'freedom' as central to different visions of the future university is the publication of 'debate books', or polemics, geared to public discussion. Between 2007 and 2010, one was published by the then minister, and others by those trying to rework and reassert some version of the 'traditional' values of the university.[44] *Death of the University* by Mogens Ove Madsen, a social economist from Aalborg University and then-chair of the Danish Association of Lawyers and Economists' university section, argues explicitly for rehabilitating universities in the spirit of Humboldt.[45] He addresses the book to the public and the academic trades unions' umbrella organization, which had supported the government's university reforms.[46] His definition of a Humboldt-style university is as follows:

> A university 'in a Humboldtian sense' is understood as an autonomous, economically independent institution with collegial self-management, where individual academic freedom drives free research and work for the scientific unit, and where there is research-based teaching, and not least, it is a home base for the right to publish freely.[47]

---

44 Helge Sander (ed.), *Fremtidens universiteter* (Copenhagen: Gyldendal, 2009); Ove Mogens Madsen, *Universitetets død* (Copenhagen: Frydenlund, 2009); Auken 2010; Niels Kærgård, Carl Bache, Mogens Flensted-Jensen, Peter Harder, Søren-Peter Olesen & Wewer, Ulla, *Forsknings- og ytringsfriheden på universiteterne. Forskningspolitisk årsmøde 2007* (Copenhagen: The Royal Danish Academy of Sciences and Letters, 2007); Claus Emmeche & Jan Faye (eds.), *Hvad er forskning. Normer, videnskab og samfund* (Frederiksberg: Nyt fra Samfundsvidenskaberne, 2010), http://www.nfsv.dk/Visning-af-titel.403.0.html?&cHash=95856c78cf&ean=9788776830229.

45 Madsen 2009, 7.

46 Ibid. 69.

47 Ibid. 9.

Arguing that the Humboldtian university is 'in free fall', Madsen relates how 'every single part of the Humboldtian definition is being eroded' by changes to research, teaching, collegial self-management, and personnel management.[48] These developments, he argues, and the government's new steering model, are eroding research freedoms, the possibility of producing impartial research, and public trust in research results.[49] As a result, 'The university, instead of having a value in itself, is increasingly assuming an instrumental value for industrial interests in the surrounding society'.[50] While these are very strong arguments, from the point of view of the analysis advanced here there are two weaknesses. First, the book does not question why and how the government uses the word 'freedom': it is as if 'freedom' is self-explanatory and belongs to the Humboldt model. This means that the book does not expose or contest its shifting meaning. Second, *Death of the University* argues for the university to be revived in the spirit of Humboldt, but it does not explain why that is relevant for the contemporary context, or the kind of world a university based on a Humboldtian concept of freedom would set out to create.

A second debate book, also referring to death—*Hjernedød. Til forsvar for det borgerlige universitet* ('Brain death. In defence of the "*Borgerlig*" university')—is by Sune Auken, a lecturer in Danish literature, public intellectual, and active member of the Dansk Magisterforening academic union.[51] Auken draws on the same cluster of concepts around the key-word 'freedom' as Madsen, but he labels them mainstream (*borgerlig*) rather than Humboldtian. He is trying to present research freedom not as a tradition or as even as an individual right, but as a current and future necessity for society. Auken tries to awaken ordinary citizens (*borgere*) to the need to defend the conventional (*borgerlig*) university against the impact of the policies of the right-wing (*borgerlig*) government and reassert the university's trans-political role in informing and critiquing mainstream (*borgerlig*) society.[52] Current policies narrow research and education to industrial needs, whereas ordinary people look for new knowledge on a wide range of topics. Universities need the conditions of research freedom '*to follow truth without being hemmed in by other constraints than those the research process itself imposes*'.[53]

48 Ibid. 10.
49 Ibid. 12.
50 Ibid. 13.
51 Auken 2010.
52 Ibid. 11.
53 Ibid. 46, original italics.

He distinguishes research freedom from freedom of expression because the former requires economic resources. Society invests these resources because it needs the possibility to search after truth, to question truths put forward by special interests, and to generate knowledge that interests citizens rather than politicians—and which, unpredictably, might be useful in 15 or 100 years' time. He then goes on to detail how the government's changes to the funding and steering of universities constrain research freedom. He concludes that this top-heavy steering by the politicians and university managers makes them deaf to the objections of university staff and students. As a result, the minister 'believes that his absurd system of "measuring" universities' activities actually measures universities' performance' and that they 'are on the way to becoming world-class'.[54] In contrast, the title's pun conveys Auken's view that the measures are an 'absurd stupidity' that will leave the universities brain dead.

Both books operate within the narrow Danish framing of research freedom, not the international recommendations' wider concept of academic freedom. Madsen treats research freedom as an individual right, and while neither treat it as a collective right to be exercised as a social duty, Auken, rarely for Denmark, argues that it is a social investment and a 'necessity from society's point of view'.[55] Both books advance similarly angry critiques of government control, but neither explore how and why the system they find so constraining and destructive can be called by the government 'making universities free'.

It is difficult to assess what impact these books had on the public or politicians—apart from a marked lack of reaction from the latter. One sign of this is the Sander's anthology.[56] He included contributions from the leading politician (Jesper Langballe) who argued for the values of the 'Humboldt' university, and from one of his leading academic opponents (Peter Harder), but in the introduction, Sander ignored them completely, and instead used the space to rehearse his claims to have put in place the foundations for universities to play their crucial role in the global knowledge economy, with 'knowledge as Denmark's trade good number 1' and universities as 'the pivotal point in the meeting with the global'.[57] Even though he avoids the motto of the reforms, 'from idea to invoice', which provoked widespread academic resistance, he presents his policies using a string of other catch phrases: 'freedom along with responsibility', stronger collaboration with 'globalised research', and students' transition 'from book to employment'. His interest is in pushing the university

54 Ibid. 44.

55 Ibid. 47ff.

56 Sander 2009.

57 Ibid. 14, 26.

in the direction of the global knowledge economy without making any concession to his critics or any appeal to Humboldtian values: the key-word 'freedom' is the only hinge between the two discourses, and he neither spells out what he means by it nor engages with his critics' ways of conceiving of freedom. As with all effective symbols, 'freedom' is attractive to diverse audiences as long as it eludes definition, leaving everyone to read their own meaning into it.

It was very rare for Sander when minister to actually meet his critics in public and for there to be a confrontation between his idea of freedom in the government-steered market model of the university and academics' concepts of research freedom. One occasion was the annual political meeting of the prestigious Royal Danish Academy of Science and Letters in 2007 on the theme 'Freedom of Research and Freedom of Expression at University'. Six leading professors prepared a report, and the minister was invited to reply to their criticisms. In their report, they set out the following freedoms: the freedom to pose questions, even against orthodoxies even when there are strong interests and feelings attached to the authorities' established thinking; the freedom to choose research topics, and decide on the material to be collected and the methods to be used; the freedom to present hypotheses, results, and reasoning in public; and the right to criticize one's own organization and others in the organization.[58] The first two refer to the narrow Danish meaning of 'research freedom' and the second two to the constitutional right of 'freedom of expression': all are framed as individual rights. The report gave numerous anonymous examples of how researchers felt their freedoms were being curtailed by strong leaders, university strategies, and the move to competitive research funding and 'internationalization' (the demand to publish in English rather than for a Danish audience). Instead, they wanted a legal framework and the organizational and working conditions that would give researchers 'real freedom of action', arguing that many social developments start from free researchers' unconstrained and partly unsystematic pursuit of ideas, and only an environment of constructive criticism stops society from fossilizing.[59] Like the other debate books, this reiterated the narrow Danish discourse about individual freedom of expression and freedom to choose the topic and method of research; it did not use the wider concept of academic freedom found in international recommendations, nor did it argue that academic freedom exists to be exercised, not primarily as an individual right that sometimes has social by-products, but as a collectively protected duty to society.

---

58 Kærgård et al. 2007.

59 Ibid. 31.

At the meeting, which I attended, the president of the Royal Academy addressed the minister, setting out the points in the report and asserting the academics' own ideas of freedom: of an individual right to pursue knowledge independently wherever it led, but in an atmosphere of respect for one another's ideas. The president, like the report, did not engage with the minister's ideas of freedom. The minister denied any reduction in freedom, and responded with a rare, unscripted speech, fired by anger bordering on disdain. He advanced many of the arguments later published in his book. Freedom of expression and research freedom, which he treated as quality controls, are commitments in law and should be protected.

One speaker after another made strenuous efforts to assert their ideas of freedom. They criticized the government for saying the policy was increasing research freedom, whereas in their experience it was being curtailed. They warned of the enormous level of dissatisfaction—later evidenced by the petition to Parliament, which the minister also ignored. What they did not do in their report and in their questions to the minister was explore what he meant by freedom. They did not expose the different meanings of 'freedom', so could not confront him with the incompatibility between their idea of research freedom and his idea of government-steered freedom, or contest the government's shift in the meaning of freedom. It was as if they heard, or 'misrecognised' the government's use of 'freedom' as their own,[60] and as a result the minister continued using 'freedom' in one way and the academics in another. They batted the word back and forth across the vast polished table in the centre of the hallowed room like a ping-pong ball.

Conclusion

If the OECD's initial move towards the market model of the university depended on simultaneously referring to the 'Humboldtian' or traditional model so as to sow confusion and mobilize popular consent for a project in the interests of industrial capital, by the end of this story, in the encounter between a government minister and leading academics at the Royal Danish Academy of Science and Letters, this dual move, stepping towards a marketized university but simultaneously signalling to those opposed to it—in Hall's terms, a double shuffle—turned on just one word: freedom. When the minister talked

60 Susan Wright, 'Processes of Social Transformation: An Anthropology of English Higher Education Policy', in J. Krejsler, N. Kryger & J Milner (eds.), *Pædagogisk Antropologi – et Fag i Tilblivelse* (København: Danmarks Pædagogiske Universitets Forlag 2005).

of 'setting universities free', giving universities 'freedom with responsibility' and 'freedom to manage', and promising 'increased degrees of freedom', his discourse reflected ideas of state control through freedom; that universities, along with other public institutions, should be given the legal status of a 'free' agent, responsible for negotiating their own relations and solvency in an increasingly marketized world. However, the Danish version of governmentality combined that liberal vision with tight ministerial and interventionist political controls exerted through detailed contracts, performance indicators, and funding mechanisms that made clear what performance was required and how the government would judge whether the university (for which read university leaders) had used its freedom in the required responsible way. In this dirigiste form of controlled freedom, universities were to concentrate their research resources and employment-focused teaching on Denmark's strategic needs in a knowledge economy; other areas of research or a more critical approach to education would be an irresponsible waste of public resources. In this discourse, research freedom was equated with peer review, a form of self-managed quality control that helped ensure value for money.

The meaning of freedom in the government discourse is conveyed by its cluster of associated terms—responsibility, strategy, knowledge, economy, enterprise, contract, performance, accountability—but the academic opposition signally failed to read the code. When government documents referred to freedom, it seems that academics latched onto the word, and believed it meant that the government shared their language. In their terms, research freedom meant the individual right of academics to choose their own research topics and research methods, and they linked freedom in a semantic cluster with research, teaching, debate, critique, ethics, autonomy, and democracy. They were puzzled and frustrated by the government's strategic funding plans and the minister's seemingly wilful inability to see how his reforms 'setting universities free' were incompatible with their 'research freedom'. The academic critics were caught out, attracted by the government's references to freedom—which it then studiously ignored. The academics did not analyse the two meanings of freedom in play, or effectively set out the differences and incompatibilities between them. By failing to expose and challenge the dominant meaning of 'freedom', or its appropriateness for a university, academics have been unable to decouple their discourse from its role as a subordinate and supportive partner in the government's double shuffle.

This decoupling would be a first step in an academic reworking of the 'Humboldtian' discourse in order to advance it as an alternative view of the role of universities in society. The discourse of Danish academic critics rarely makes a strong social argument; its focus is on the rights of individuals, and it relies mainly on the right of all Danish citizens to 'freedom of expression', in preference to upholding the particular legal right of all academics and the university as a whole to 'research freedom'. Nor does the Danish academic discourse make full use of international recommendations, which state that preserving 'academic freedom' as a critical space distanced from the demands of government and the economy is the academics' collective responsibility and duty on behalf of society.

By not exposing the government's meanings of university freedom, not decoupling 'Humboldt' from the government's grasp on its subordinate and supportive role in the double shuffle, and not advancing a strong alternative vision of the role of the 'freed' university in and for society, academic critics have failed to halt the Danish government's momentum towards a university steered through freedom in the service of the knowledge economy. Vociferous academic resistance has been sustained, but has not effectively dislodged the political dominance of the minister's line on freedom. Among politicians at least, that discourse had become so dominant and so taken for granted as a way to see the world that when there was a change to a Social Democrat-led coalition government, the new minister, in his first interview with the academic press, said his aim was to make universities more free.[61] I have not found any letters to the editor or comments questioning what exactly he meant by 'freedom'.

---

61 Morten Østergaard, '5 skarpe til ministeren' *Magisterbaldet,* 16 October 2011, 8.

CHAPTER 9

Philosophy, Freedom, and the Task of the University

Reflections on Humboldt's Legacy

Hans Ruin

Philosophy and Humboldt's University Ideal

In Wilhelm von Humboldt's programmatic text for the new university of Berlin from 1809, the university is presented as a unique institution in which human beings could reach their highest potential in a shared unbound search for truth.[1] Such a place should be established and maintained by the state and provided for through public funding, yet the state should interfere as little as possible in its internal affairs; by allowing the university to operate freely, the state itself would grow in freedom, and, ideally, the university then becomes the very place in which the free state can expect to find its own idea embodied. At the time of the creation of the University of Berlin, the university in general as a cultural institution was not always looked upon with favourable eyes. It was partly through Humboldt's commitment to the university as an idea, and before him Kant, Schiller, and Fichte, that once again it could become an object of idealistic hopes. Following the efforts of this generation, the modern idea of an institution dedicated to the infinite growth of knowledge, and the fostering of citizenship and social bonds, emerged as the embodiment of the spirit of the Enlightenment. This was also the time of the rapid secularization of Western culture. The modern university, together with the national museums, henceforth become intellectual focal points, through which these modern societies gained a new sense of direction, partly at the expense of the Church.[2]

1 'Über die innere und Äussere Organisation der höheren wissenschaftlichen Anstalten in Berlin', in *Werke*, iv (Darmstadt: Wissenschaftliche Buchgesellschaft, 1982).

2 This account is not meant to imply that it was primarily the idealistic discourses of Enlightenment philosophers that led to the creation of the modern university as an institution. As has been shown by university historians, the process leading up to the formation of the modern university in Germany at the end of the eighteenth century was a complex web of interrelated social, political, and economic events. See, for example, the excellent survey in R.D. Anderson, *European Universities from the Enlightenment to 1914* (Oxford: OUP, 2004),

 | DOI 10.1163/9789004271944_011

The themes that have left their mark on the debate about the university as an institution are intimately and inextricably tied to Enlightenment philosophy and its central concepts—reason, knowledge, truth, critique, and *Bildung*—all of which are gathered in the idea of *freedom*. The university is a place that should make possible the growth of knowledge, in a communal spirit of critical analysis, free of all presuppositions; it is in this way that the university should embody and promote the very idea of freedom. In his memorandum, Humboldt writes of how this way of life should rest on the freedom of teachers as well as students: freedom of learning, guided by free teachers. In his programmatic text, Humboldt does not write specifically of the role of philosophy, except in passing. Yet central to this whole thought-complex was a belief that philosophy was the quintessence of *Bildung*. When, ten years earlier, Kant considered the organization of the university in 'The Conflict of the Faculties', he had singled out and defended the significance of philosophy as a guiding principle for the ideal and the practice of the university.[3] In terms of student numbers, the leading faculties—the theological, legal, and medical faculties—would predominate, primarily devoted as they were to the training of civil servants, and thus to the task of serving a well-functioning society. The philosophical faculty, meanwhile, worked with a more diffuse, yet all the more fundamental idea—for according to Kant, it was only in philosophical pursuits that a more *free* and exploratory knowledge process can take hold, a mode of research that is free from assumptions and that makes possible a mutual critique.

From the viewpoint of society as a whole, it was precisely the rather diffuse character of philosophy that justified its position as a subordinate, 'lower' faculty. Yet the fact that it alone carried and maintained this ideal compelled Kant to argue, in a revolutionary turn, that in fact, and from another perspective, philosophy must be seen as the highest-ranking faculty of all. Only if this spirit permeated the philosophical faculty could it then govern the whole, so that the university in its entirety could fulfil its task. In other words, even if a higher-education institution necessarily took as its main task the training of individuals to serve in the different institutions of society—doctors, economists, lawyers, engineers, priests and teachers—in its organization it should also be permeated by a critical spirit springing from the free formation of knowledge. This meant for Kant that the university had to be permeated by the

---

which also shows how Humboldt's ideals did not definitively influence the development of the European university system generally until much later in the nineteenth century, and then partly as the result of a new wave of secularization.

3 Immanuel Kant, *Der Streit der Fakultäten in drei Abschnitten* (Hamburg: Felix Meiner, 2005).

philosophical ethos. In a university faithful to its task, philosophy would have a central and fundamental role as the concentrated expression of the spirit that constituted the very essence of the university. In 1789, as part of his inaugural lecture as a professor, Schiller too had distinguished the 'bread-fed scholar' from the 'philosopher'. Whereas the former only sought knowledge for immediate gain, the latter represented a love of knowledge and of wisdom, in a free community of questioning. It is this philosophical attitude that, in the final instance, represents what life within the university should be.

In post-Kantian philosophy, this idea of philosophy, and its connection to the overall task of the university and the guiding idea of *Bildung*, was to undergo a series of variations and transformations. The task here is to recall some of the stations along its way, in order to bring us up to the present, where the role and position of philosophy in the modern university is increasingly questioned. Ultimately, I will place Humboldt's programme in relation to the establishment of a new, state-funded university institution in the making, one with which I have been closely involved over the last decade, Södertörn University in Stockholm.

In a commentary from 1980 on 'Conflict of the Faculties', Derrida echoes Kant's statement that without philosophy, there is no university.[4] But there he also speaks of the open-ended nature of academic responsibility, and the ambiguity of the ideal of academic freedom. Every act of autonomous initiation, for example of a philosophical institution, cannot in itself rest on the secure foundation of what it means to be philosophical; instead, it will require taking responsibility for the meaning of what such an institution might be. In May 2000 this comment was actualized again when Derrida participated in a seminar devoted to his work at the newly founded Södertörn University in the southern suburbs of Stockholm. One reason why he accepted the invitation from the philosophy department at the then entirely unknown institution was that he had a particular interest in the possibilities that the founding of an institution offered. In his introductory remarks he spoke of the importance of these founding moments in the history of academic institutions, precisely as a way of exploring the possible, and he encouraged the work that had just been initiated. The following remarks should also be read as an attempt in part to respond to the challenge that he posed to those present on that memorable occasion.[5]

---

4 Jacques Derrida, 'Mochlos ou le conflit des facultés', in *Du droit à la philosophie* (Paris: Galilée, 1990), 398–438, in particular 428 and 435.

5 The whole seminar was taped and published in Swedish translation by Nicholas Smith as an appendix to *Lagens kraft* (Stehag: Symposion, 2005), the Swedish translation of *Force de loi*.

Threats Facing Philosophy in the Academy

Today, philosophy, as an academic discipline, is threatened from the outside, but also from within itself. From the outside it is threatened by the simple fact that university managers are always on the hunt for ways of cutting back and making the organization more efficient, and philosophy is often seen as a redundant discipline, with questionable relevance for the academic institution as a whole. Yet even from within, university-based philosophy is undermining itself by adapting to the modern research paradigm, and becoming increasingly specialized, focusing on small solvable problems and publishing mostly in specialized journals. In the short term this gives certain benefits in the form of increased publishing rates, but in the long run this way of relating to the task and responsibility of philosophy inevitably prepares for its gradual demise. This is a 'normalization' of philosophy as a discipline. What for Kant and Humboldt's generation was self-evident—namely that a university needs philosophy among its disciplines, and that its attitude and ethos capture something essential to the scholarly practice as a whole—has nowadays been eroded to the extent that it is no longer clear that philosophy has a given place within the modern university, let alone a clearly defined task.

Moreover, what Kant took to be two sides of philosophical activity—the practical and the theoretical—have in the Swedish system developed into separate research fields, with ever weaker connections to a shared agenda. In the attempt to define itself primarily in terms of, say, a certain method, mainstream analytical philosophy has often distanced itself from the task of critically and productively appropriating its own past. Amongst many elements of analytical philosophy, the history of philosophy had, in principle, something to be forsaken—or at best something to possess general knowledge about. Either way, the history of philosophy did not appear as having any significant bearing on true philosophical work, which instead had to do with the application of chosen methods to contemporary tasks. This division of philosophy into its historical and systematic components was reflected in the way the subject was taught, with courses on method kept distinct from purely historical courses, and where the very conditions for philosophical thought were not thematized. Put differently, the hermeneutico-historical dimension of thought was not recognized as a philosophical problem in itself, or as an experience that demanded philosophical thematization.

Within the modern university, all disciplines are moving towards greater specialization, following what appears to be an inevitable logic of scholarly endeavour. According to this line, the history of philosophy then becomes a task for the discipline of the history of ideas, which can isolate this as a domain to investigate with specific methods. Philosophy can then devote itself to one of its specializations, such as formal semantics, cognitive processes, or some strand of applied professional ethics. This is how the situation presents itself in most places, where the overarching structure of research leads individual researchers to move in the same direction too. A tendency towards specialization appears both unavoidable and well motivated. But what does it mean for a subject such as philosophy, which in a unique way can still be said to carry a responsibility for the general—that is to say, to maintain not just one or another strand of knowing, but to articulate the task of knowing as such, of the *university* as such? Admittedly, such a question can appear both presumptuous and untimely—and yet, by the same token, it is impossible to dismiss, particularly for those who today wish to defend the existence of the subject of philosophy in any institution for higher education.

This is impossible to avoid, simply because it is no longer obvious what philosophy is, and what its place is, if any, within the university. For what kind of knowledge does philosophy represent? How is its knowledge construed? And how might this knowledge be transmitted, passed on to posterity and received by future generations? As long as the subject only endures within the framework of an established institution, these difficult questions can be avoided. But in the founding moment, in the moment of creating a new institution, such problems are encountered directly, compelling us to make certain choices.

Experiences from Establishing a New Department of Philosophy

The way in which philosophy came to be introduced at Södertörn University in Stockholm is in this case an example worth reflecting upon, given the institutional setting, the processes involved, and the decisions that needed to be taken. To begin with, it was perhaps symptomatic that when it was first founded in the mid 1990s, it was not clear whether Södertörn University (then Södertörn College) would even have a philosophy department. Malmö University, which shares a similar history, chose to create a modern university structure without philosophy, and the same has been true of many of the new universities in Sweden. In Malmö, as elsewhere, there are people with doctorates in philosophy, but they are normally housed within different kinds of cross-disciplinary humanities- or communications-oriented departments.

Among the first disciplines to be established at Södertörn University was instead the history of ideas, which was then asked to organize some of the inter-disciplinary and inter-faculty courses of the kind that would previously have been taught by philosophers. It was thus a 'historical' conception of philosophy that was accepted, but as the discipline specialized in exploring the history of philosophy, not philosophy in a strict sense. One reason for this choice was the fact that philosophy was normally researched and taught at the mainstream institutions in a way thought irrelevant for what the organizers wanted to construct, which was a more interdisciplinary environment. So at this particular point it was demonstrated that standard analytical philosophy, because of its specializations and its reluctance to engage in interdisciplinary work, was no longer seen as relevant for a modern institution of higher education. Two hundred years after the great university reform of Humboldt and his generation, the idea of university philosophy had lost not only its appeal, but also its aura of somehow carrying the spirit of university culture and learning. And here the examples of Södertörn University and Malmö University, and indeed of several other small, modern universities in Sweden, are not singular cases, but point to a larger trend in modern university policies. When philosophers today are invited to create centres of learning and research, it is more likely to be in the fields of cognitive science or applied ethics than in philosophy in sense of the Romantic university of Humboldt and his generation.

That a philosophy department was established at Södertörn University at all was partly the result of a series of coincidences. Through contacts with some of the university administrators, I was told quite informally in 1998 that there could be an opening to discuss the possibility of some kind of philosophy course. I wrote a proposal for a programme that contained an analysis of the situation—largely the same as the one rehearsed here—and emphasized that it would be a programme with a more 'Continental philosophical' profile, one that would not build on the division between practical and theoretical philosophy, and that would combine the systematic and historical approaches to the subject. Subsequently, I was invited to discuss my thoughts with the Dean of Humanities, specifically about how the submitted proposal might be translated into a course. At this stage there was still no principal decision to establish a philosophy department. The conditions were stated quite bluntly in financial terms: the university as a whole was expanding rapidly, and needed to grow in student numbers. Instead of asking me to develop a programme, I was offered the possibility of giving a one-semester course. If it attracted enough students to pay for its costs, it could continue. In short, there was no obvious academic responsibility on the part of the university's senior management; no sense that the institution as a whole had a need for whatever philosophy could provide, apart from its capacity to generate a certain number of students: a mandate to give a one-term course was what was on offer. Furthermore, the Dean proposed that this course be based at the Department of the History of Ideas. The stated reason for this particular stipulation was a suspicion that the situation of philosophy in Sweden was not necessarily conducive to a 'Continental philosophical' approach, and such a department might find it difficult to defend its place. Most likely there were also other motives, having to do with potential academic rivalry and competition between disciplines.

Taken together, these events invite a few reflections. The modern university today—and this applies also to state-financed economies—is largely run like a business enterprise, with the demands of economic profitability to the fore. Additionally, the university has inherited a set of lingering practices from company culture: consumer and producer meet in the production of goods to be exchanged. In the state-funded university system, the state is both the producer and the consumer at one and the same time. It orders a certain number of products from the producer (a branch of the state itself), which is then compensated for the delivered goods. In the context of the university, what the state orders are completed courses of study. It is the students who conduct the work by themselves, while the university offers the teaching and the approval of the students' performance through accreditation: this is the product that is delivered to the state in exchange for a certain payment. These, in brief, are the financial conditions for higher education in the modern business economy of the European university.

Within this system there is no essential difference between the various disciplines. All are subjected to the same conditions—from tourism and sports management (two recently created university disciplines) to philosophy. This situation means that the universities are competing in a market where the task amounts to creating attractive course programmes that can generate student results. It is a system that leaves the field free to develop pretty much any education programme as long as it meets minimal academic standards. It offers no directives on the desirability of particular course contents, and no particular sense of hierarchy.

All of this leads to a somewhat peculiar situation, especially for the humanities. On the one hand, the whole arrangement appears entirely nihilistic: there is no longer any overarching idea about what or why a certain education programme should be privileged above any other. On the other hand, the arrangement presents certain opportunities. The financial control that the state exercises over these activities is total—universities simply have to meet the standards of good financial performance and accountability. But with regards to ideas, the neo-liberal state has largely renounced its responsibility for content, let alone any overarching aim. It is true, of course, that the state gives its educational institutions certain instructions and demands as to the general output of students in designated areas, such as medicine, the social sciences, the natural sciences, and so on, based on general estimates of what the market will need in terms of skilled workers in the different areas. Courtesy of the Swedish National Agency for Higher Education (*Högskoleverket*) the intra-scientific situation is also regularly reviewed in order to ensure that there is a satisfactory quantity of teachers with a doctoral degrees, a reasonable gender balance, and that students receive the required number of hours' teaching for the economic compensation paid by the state. However, when it comes down to the details, the freedom to shape the content of the education is delegated to vice-chancellors, deans, and heads of departments.

Often this means that decisions related to the actual thematic content of courses is, at least nominally, delegated to the lecturers themselves and the boards of the different disciplines. At least, this has been the experience at Södertörn University. As long as a discipline is considered financially viable, it has, in principle, full autonomy over the contents of its programmes.[6] And as long as it can show that it 'produces' enough results, it will be given a mandate to continue, and to develop new courses. Within five years, the subject of philosophy at Södertörn grew from an experimental, introductory course to a fully developed department with courses given from basic undergraduate degrees to Master's degrees—along with a postgraduate research student programme, first developed in collaboration with Stockholm University and later with Uppsala University.[7]

There is also reason to reflect on the original proposal to incorporate philosophy into the Department of the History of Ideas. History of ideas, or intellectual history as it is sometimes also called, has a strong position in the Swedish university system. One of the reasons for this is the way in which philosophy developed in the post-war period: in many places it took on an ever more defensive methodology, immersing itself in increasingly technical questions. This meant that the history of ideas provided a platform for many researchers interested in critically interpreting the classic works of philosophy. It is no coincidence that some of the most prominent historians of philosophy

---

6 It has to be added that the conditions under which the humanities at Södertörn could develop in relatively free and open way was unusually congenial due to the engagement shown from the start by the vice-chancellor Per Thullberg, as well as his successor Ingela Josefsson, both of whom had a background in the humanities and who were explicitly guided by the Humboldtian ideal. Currently the school is undergoing a sharp turn toward managerial control-politics, following a so called 'autonomy' reform introduced by the liberal-conservative government in 2011.

7 That all this did not happen by itself, but only by way of a great effort by a small group of people, is self-evident; however, for reasons of space I will not dwell on the history of the more personal–practical side of establishing the discipline.

have taken up residence in departments of the history of ideas, and not in philosophy.[8] An important, if not crucial, question for the future of the discipline of philosophy concerns how its own history should be treated. Among the historians of ideas, philosophy is understood to be one product of culture among many, and as such should be studied with the methods at their disposal, like any other cultural artefact. There is, however, an inbuilt limitation to this contextual approach to philosophy. In historicizing and contextualizing a text, the historian of ideas tends to shy away from the more difficult and nebulous problem of articulating and engaging with its relevance and possible 'actualization'—in short, of allowing its content to reach into the here and now.

On the Future Role of Philosophy in a Neo-Liberal University

It is essential for a living intellectual culture that there remains a space within this culture where the past is kept 'open', where tradition is not only studied with the tools of the historian, but also challenged and brought to life. This is crucial to generating future orientations. The present should not only approach the past in order to have its methods and standards confirmed, but also to experience a form of estrangement from both itself and its here and now. It is all about being open to the experience of a *crisis* in the present. The standard strategy towards one's own 'contemporaneity' is to obscure it, since the now usually thinks that it knows the now, since it is the now. But what an open and critical approach to the past can teach us is how best to destabilize this conception. If *Bildung* is still to have any meaning for us, it should also be seen in its ability to generate the experience of a liberating estrangement from one's own sense of self and presence, and thus of a profoundly philosophical experience of the present as mediated through itself. For Humboldt, this experience was something that he elicited primarily from his extensive work on translation, and how the encounter between languages served to recall what was foreign also in what is one's own.[9]

That this should be an important aspect of philosophy is, however, something that few of its representatives today wish to subscribe to. Recently, I participated in a conference at the Swedish Research Council (*Vetenskapsrådet*)

8 It is this situation that also prompted investment in a post in the history of philosophy at Uppsala University in the early 2000s, and effected some change in recent years.

9 For Humboldt's understanding of *Bildung* and the problem of translation, see my 'Att översätta sig in i det främmande—idéer för ett samtida bildningsbegrepp', in M. Lindh et al. (eds.), *Från Högskolan i Borås till Humboldt* (Borås: Högskolan i Borås, 2011), 45–54.

that considered new methods for the evaluation of research quality in the humanities in general. On this particular occasion, philosophy had been picked out as a case in point. All the philosophy departments in Sweden were represented. In this circle there was a strong consensus that philosophy had made huge advances in the twentieth century, and that we were part of an innovative phase where many of the old problems had found their solution; the general agreement was that this showed itself not least in the frequency with which Swedish researchers published in Anglo-American journals. The picture painted of Swedish philosophy by its representatives was that of a successful 'normal science' (in Kuhn's sense) with a contemporary, international research horizon. In the light of this self-image—and before the coming system for scholarly qualifications and research funding—it remains a challenge to try to keep the space open for a more critical, hermeneutic form of philosophical thinking. For, on the contrary, it is within such a tradition that the self-assertion of the now can be tested in light of a critical appropriation of the past. Ultimately, and despite the self-confidence radiated by its representatives during the meeting, it is the very position and legitimacy of philosophy within the university system that are at stake.

This goal is also indirectly connected to the question of how one can work today to uphold something like the ideal of the 'freedom to learn' and the 'freedom to teach' in Humboldt's sense. After all, there is a growing requirement for ever greater detail in the planning of university teaching, in a way that all the more seems to radically limit the possibilities for individual university teachers to continue to create and develop their own courses. The standardizations brought by the Bologna Process are a driving force here. In tandem with increasingly review-steered research financing, the present set of conditions point towards the curtailment of the university teachers' knowledge development, and thus go against the grain of Humboldt's original hopes for academic freedom. Here it is important to cultivate a certain type of passive resistance to the negative consequences of standardization. When setting the goals for Södertörn University's new philosophy department, one ambition was to create an undergraduate programme that combined the demands of a basic training in philosophy (in terms of a knowledge of the disciplines and subdisciplines, methods, writers, and so on) with an opportunity for the lecturers to integrate their own research interests into the syllabus. By designing the courses as openly and non-specifically as the formal bureaucracy permitted, it was possible to change not just the course literature but also the themes of the courses from one semester to the next. Thus no semester has been exactly like another. Before every new semester and new academic year, the lecturers are able to express their preferences for what they would like to work with, and, from there, we investigate how this can be incorporated within the reasonably porous course platforms already in place.

This strategy of preserving some degree of *Lehrfreiheit* in the Humboldtian sense has been possible by stretching the existing university regulations as far as possible. Yet, none of the two independent reviews by the Swedish National Agency for Higher Education has breathed a word against the principle of this tactic. Similarly, our experience with basically commercial university bureaucracies tells us that in reality university lecturers' freedom of action is generally far greater than most would admit to. The modern university bureaucracy is permeated by an impulse towards self-control, nourished by an illusory perception that the system can ultimately secure its own future and its own 'quality'. As a result, the academic system devotes itself increasingly to reviewing itself. What can be secured—and what should be secured—are many of the measures that the state gauges, via its national agencies' assessments of the number of hours of teaching per student, the forms of examination, the teachers' formal competence, and so on. There is nothing strange or wrong about these basic control routines; however, 'quality' in a more significant sense cannot be secured by such parameters. Quality lies hidden in best practice, and thereby presupposes the qualified skills and trained judgement of individual practitioners.

To understand how this works, one needs to be able to *think* through the systematic and conceptual nature of academic practice, both in terms of individual performance and institutional structure. Instead, what often happens today is that the intellectually shallow and commercial way in which the system is organized (not entirely without good reason) is internalized by its participants, who feel themselves to be restricted and watched. With the experience gathered over the years spent starting a new department at a modern, state-controlled institution, I would still claim that these days the threat to 'academic freedom' in the Humboldtian sense does not just come from outside; to an equal extent, it is generated from within the academy itself. In order to see this, we must also revisit the historical situation in which Humboldt's discourse was formulated, and with a critical eye if we are not to be misled by its rhetoric. There is a clear tendency for the contemplative life of thought to itself become bureaucratized, as a result of its practitioners unconsciously taking on the state's commercial definition of own activities. In the end, it is probably erroneous even to think in spatial terms of the threats facing the university today—that is, in terms of an 'inner' and an 'outer' threat. In fact, this traditional topology as articulated by Humboldt himself appears increasingly illusory, not because it was originally mistaken, but because of the transformation of society and way in which higher education and research are viewed today. When we discuss academic freedom today, we are operating within a framework that was established by the founders of the modern university, for whom that self-same 'inner' academic space was to be protected from 'outer' political and religious interference. The free and open university has long since won those battles in democratic societies: there is no interest now in controlling universities as a whole, steering them towards a particular set of politically or religiously motivated ideas (even though there are, of course, continuous demands for knowledge to be 'useful', according to shifting standards). In this period of neo-liberal policies, the basic orientation points are profitability and efficiency. It is rather the lack of vision and the absence of positive thinking about what universities are, and could be, that today define the organizational bureaucracy.

To the extent that it is reasonable to talk of a university *crisis* today, one component would be a lack of concern with the role and specificity of university culture in and for society as a whole. As an institution, the university is far from being in crisis; at least, not if one looks at its sheer size and number of employees. It was never clear from the start that 'the university' would become the kind of institution it has become, still carrying all the prestige and hopes vested in it. With the huge expansion of higher education over the last fifty years, governments all over the world have chosen to transform their universities and university colleges into halts on the path to full, mature citizenship. At least in this respect the idea of the modern university has succeeded far beyond the hopes of its original creators and defenders. Today, the majority of the working population in the more advanced economies are expected to pass through university at some point during their life's journey.

Yet, at the centre of this victory for the university as such, there is nonetheless an evident vacuum with regards to what this institution is and what it could be. It is symptomatic that it is rarely within the university itself, but rather in the independent cultural journals and external publications, that the future of the university is discussed in anything approaching a qualified manner. What is lacking within the university is a point from which the sense and essence of the university can be reflected on, and critically assessed, in a free and philosophical spirit. This role does not perhaps require philosophy to be established as a discipline or as a specific department. However, it is a task intimately bound up with what we still think of as philosophy: a certain ability to step out of oneself, to critically and systematically reflect on the point from which one thinks, speaks, and acts; to view thought in both a contemporary and a historical light; and judge oneself in the light of a different possible life.

It is a challenge to resist the bureaucratic impulse in oneself, and above all the impulse to internalize the empty economies of the system, and instead to work for continued freedom of thought.

Things will go amiss if we believe that it is possible to fight a definite enemy, as if we still lived in a pre-revolutionary times, or in a time of revolutionary dictatorship. It is no coincidence that it was the philosophy departments that were subjected to the most severe control under communist rule—not least Humboldt's own university in Berlin. For while most practically oriented disciplines (such as medicine, physics, and biology) can be pursued under political oppression, a philosophical department worth its name cannot be tolerated, for its vocation is ideally to work for freedom and truth. But when political freedoms have been secured, and the state no longer interferes in the life of its citizens, then, as long as everyone behaves well and everything is profitable, a new situation for philosophy departments arises. Now the challenge is of another kind. It is not only the philosophy department itself that is at stake, but philosophy as such, whichever the department; philosophical zones and hothouses within the academy need to be fostered, as does a thinking that works in a spirit of solidarity with the whole, and yet retains a productive distance from it.

In the end, it is only such a spirit that can nourish real resistance to the anonymous and impersonal mechanisms that will otherwise drive the university in the direction of a fully fledged system of production and of use–value maximization, neither of which is much given to self-reflection. Such a state of affairs does not happen because somebody absolutely wants it to happen, but because the very logic of the system tends to take over the participating individuals' sense of self, making them its servants. Such is the case with individual researchers, who increasingly see themselves as egoistic value-maximizers, competing for funding for their own little particular strand of science.

If, on the contrary, one makes an outer enemy out of the administrators and steering committees, meeting the state head-on as an opponent, one is likely to feed a reactive response—which again is unproductive, and in the end inauthentic. Most academic staff at some point or other take part in the review organs and processes of the system, contributing reviews, assessments, evaluations, and, ultimately, decisions. If one sees this apparatus as only an external threat, instead of accepting it as to some extent a natural dimension of the system, one will contribute to creating a self-loathing bureaucratic body that sustains with one hand what it attempts to fight with the other. Instead, we need to think clearly about the current conditions with the help of a more refined topological model, where the very idea of inner resistance to an outer enemy is replaced by a strategy of creating free spaces within and outside the structure, making it work in favour of its own possibilities, and not simply accepting it as a necessity.[10]

If, on the contrary, one gives way to the idea of a necessary or constrained mechanism, the very idea of free thinking will soon appear superfluous, a luxury we can neither afford nor need. It is in the face of this new thoughtlessness, which curiously breeds within the system, that the philosophical spirit has to seek out new forms. It has to dare to become 'untimely', in the Nietzschean sense; it has to dare to stand outside, above as well as below its present, which otherwise risks collapsing under the weight of its own self-preoccupation; and it has to dare to take the leap to think the whole, not shying away from its responsibility, but working in an unconditional way within and for this whole. Moreover, it has to cultivate attentiveness, so that in the familiar, one seeks the most different, and so in the strange one learns to see what is most proximal. It has to strive to overcome its own cynicism and despair, and to work together with others. Only in this way can something of what we still can, and ought, to associate with the freedom of the academic spirit, as once envisaged by Humboldt, be kept alive.

---

10 These reflections will perhaps call to mind the analyses by Helga Nowotny, Peter Scott, and Michael Gibbons on the implications of what they have labelled Mode 2 science. When they introduced the concept in *The New Production of Knowledge* (London: SAGE, 1994), the idea was explicitly to question earlier schematizations of how the academy and society interact. In the preface to their more recent *Re-Thinking Science: Knowledge and the Public in an Age of Uncertainty* (Cambridge: Polity, 2001), 1 they write: 'science could no longer be regarded as an autonomous space clearly demarcated from the "others" of society, culture and (more arguably) economy. Instead all these domains had become so "internally" heterogenous and "externally" interdependent, even transgressive, that they had ceased to be distinctive and distinguishable (the quotation marks are needed because "internal" and "external" are perhaps no longer valid categories)'. My reflections on the need to find new ways of orienting oneself in the academy should perhaps be read as a contribution answering the question 'What has science to offer life?' (ibid. 244) and in their appeal for a deepened reflexivity and participation on the part of academics (ibid. 262).

CHAPTER 10

Reclaiming Norms

The Value of Normative Structures for the University as Workplace and Enterprise

Ylva Hasselberg

This chapter will consider how the image of the university—the stylized picture of the university as it was, or rather, as it should have been—is used in the present discourse on science in society. The university's institutional and research norms have been discussed since the early nineteenth century; indeed, stylized images of the Humboldt university (teachers' and students' freedom of thought, the organic links between research and teaching) and of Mertonian science (open, liberal, objective, and disinterested, based on CUDOS) have long been used as tools to create legitimacy and trust in science and the university as an institution. These ideal types are invoked by many who gladly promote the commercialization of both science and teaching. As Daniel Greenberg has demonstrated, even the most devotedly capitalist of university presidents will embrace the values of open science, especially when charged with breaches of code of conduct.[1]

The power of these rhetorical tools lies in their ability to persuade the public that universities are peopled by actors motivated by a passion for knowledge and truth for truth's sake. This is, of course, a problematic idea, because, as science, technology, and society studies (STS) concluded decades ago, scientists are humans just like the rest of us, sometimes driven by greed or revenge, sometimes just doing their job. Historian of Science Steven Shapin seems to claim in his book *The scientific life*, that there is little need for the idea that scientists are of a different moral quality than the rest of us.[2] He sees the purpose of this idea mainly as a tool for boundary-work: excluding certain groups, like scientists working within industry, through questioning their moral purity.

My purpose here is to continue this discussion, and to argue that the normative aspects of science are important, not only as rhetorical tools, but in response to an emotional need for the *Sachlichkeit* (objectivity) that is the

1 Daniel Greenberg, *Science for Sale: The Perils, Rewards and Delusions of Campus Capitalism* (Chicago: University of Chicago Press, 2007), 101–126.

2 Steven Shapin, *The Scientific Life: A Moral History of a Late Modern Vocation* (Chicago: University of Chicago Press, 2008).

 | DOI 10.1163/9789004271944_012

basis of a concept of good work. *Sachlichkeit* has not only connotations of facticity but also a certain emotional colour, a strict abstention from emotional engagement: to be objective is to refrain from emotional engagement, and this can be a quality of a subject or a personal attribute. Without this emotional link to *Sachlichkeit*, science in itself lacks meaning for those that engage in it. A definition of first-rate work, and the norms connected to this definition, is central to the identity of any professional group. It is for this reason that we cannot leave question of university norms to celebratory speeches, but must strive to reclaim it for ourselves, for the academic community. I see it as emancipatory to try to draw a distinction between how norms are used politically, and what they do in terms of reproducing standards of good work. This is my main concern in this chapter.

On Norms and Interest

The professionalization of scholarship is a central concern here, and reflecting on this process throws light on both on the university as an institution and on professionalism—its characteristics and meaning. The professionalization of scholarship is a question of specialization as well as the development of rules and routines aimed to create a knowledge that is qualitatively different from other types of knowledge. In looking for the driving force in this process, I, like many others before me, look primarily to the development of scientific method, in the sense of the rules of how to conduct science: how to ask questions, how to conduct investigations, and how to write text. Conducting science in a manner informed by a specific methodology is possibly the core of academic professionalization, and far more central than the development of theoretical paradigms or the organization of science. This core has often been confused with the development of technical competence: professionalization is something else, and requires more than the development and acquisition of technical competence. Technical competence is embedded in ideas of how thing *should* be done, it is thus value-based. Mistaking technicalization and the parallel growth of a cadre of salaried technicians for professionalization would be unfortunate in the extreme.

Modern science has its particular apostles. The fact that science has developed in a historical process where professionalism was created and advocated by specific historical actors is a banal fact, but not an innocent fact. The fact that professionalization is a social process is not only a cornerstone in the theories that hold it as symbiotic with modernization, but also in the notion that professionalization is socially constructed in another sense—not as 'developed in a historical process', but as 'a hoax aimed to benefit a particular interest'. According such a view, the value base of professionalism is in reality non-existent, and can be reduced to social mechanisms that exist to promote this interest.[3] Professional praxis and professional institutions are thus seen as a mixture of hocus-pocus, habit, and deliberate exclusion.

There is a growing suspicion in society that phenomena such as peer review, methodology, and disciplinary boundaries, which create distinctions on the basis of value-based judgements, are of little or no worth in advancing knowledge. What then seems to be acceptable and understandable in the legacy of professionalism is rather what seems to liberate knowledge from the subjectivity and values of professionals—that is, rules, routines, and techniques. These (for example, 'evidence-based science') form the core of a counter-professionalism (sometimes called the new professionalism), which holds that the professional, in this case the scientist, often is and *should be* a salaried technician, not a person devoted to a calling. Note the normative aspect of the new professionalism. At the end of the day, it seems hard to escape the normative aspects of professionalism, even though they can be hidden under a cloak of standardization. It is deemed professional to follow 'best laboratory practice';[4] it is deemed less professional for professionals to follow their own judgement and do things in an order that is motivated by a complex context: a norm of non-normativity and of non-judgement seems to be inherent in the new professionalism.

This brings us back to the apostles of modern scholarship. The claim is certainly correct that the status of science in society and the research-based professionalism of the modern university are the results of a conscious rhetoric intended to accomplish just this. The most intensive efforts were seen in the inter-war period, when Max Weber gave the speech that became 'Science as a vocation' (1919), Karl Popper wrote *The Logic of Scientific Discovery*, and Robert

3 Magali Sarfatti Larson, *The Rise of Professionalism: A Sociological Analysis* (Berkeley & Los Angeles: University of California Press, 1977).

4 According to the concept of 'best practice', it is possible to discern a superior method of undertaking certain actions, on the basis of a cumulative advancement of knowledge, and that a manual can be constructed that reflects this knowledge-base. The individual professional is thus relieved of the right and duty to make decisions regarding practice in every single case. Following the manual is seen as a guarantee for fewer mistakes and better results. There are also economic aspects of best practice. Its purpose is not only to improve quality, but also to increase efficiency. See, for example, Center for Disease Control and Prevention, *Laboratory Medicine Best Practices. Developing an Evidence-Based Review and Evaluation process. Final Technical Report 2007: Phase 1* (Atlanta: US Department of Health and Human Services, 2008).

Merton published his doctoral thesis (1938) and his first article on the value system of science, 'Puritanism, pietism and science' (1936). It is no coincidence that the early twentieth century was also the period when the legacy of the Humboldt university began to be seen as an emblem of a particular way of organizing research. It was a period of external professionalization (in its organization and creation of legitimacy) as much as internal professionalization (in the growth and specialization of knowledge meant to benefit scholarship). The historical actors who argued publicly for autonomous science also led the way in intra-disciplinary professionalization in their depoliticizing efforts and their methodological and theoretical work.

Steven Shapin, in *The Scientific Life*, correctly notes that the inter-war period was a time when beliefs about the motives and driving forces of individual scientists shifted: the nineteenth-century concept of the scientist as a particular sort of character, possessed of a different moral quality than the rest of society, gave way to the idea of the scientist as the moral equal of the man in the street. In early modern Europe, the social and moral standing of the scientist had been essential components in the process of judging the validity of their scientific results.[5] In the Victorian context, which gave birth to the modern research university, this moral quality was still uncontested. Before becoming a paid professional, the researcher was an amateur, per definition a person who did not have a financial interest in scientific activity. Science was the pursuit of gentlemen. To be devoted to this non-profitable activity, the scholar had to be in possession of certain moral qualities. As Eli Heckscher, the founding father of Swedish economic history, put it, 'a scientist must have character'—something he had picked up from his teacher David Davidson.[6] The stress on character, meaning individual moral strength developed through active self-improvement and self-discipline, was typical of the late Victorian period, and it was also, as Swedish historian David Tjeder has demonstrated, central in the then dominant concept of masculinity.[7]

It is uncontroversial to argue that both late nineteenth-century professionalism and the modern research university were coloured by the Victorian claim of the significance of moral fibre and character, and that later, during the

5 Steven Shapin, *A Social History of Truth: Civility and Science in Seventeenth-century England* (Chicago: University of Chicago Press, 1994).

6 Ylva Hasselberg, *Industrisamhällets förkunnare: Eli Heckscher, Arthur Montgomery, Bertil Boëthius och svensk ekonomisk historia 1920–1950* (Hedemora/Möklinta: Gidlunds förlag, 2007), 78–79. Translation by the author.

7 David Tjeder, *The Power of Character: Middle-class Masculinities 1800–1900* (Stockholm: Stockholms universitet, 2003).

inter-war period, a new view began to emerge, and it became more common to assume that the scientist did not differ from other people in terms of moral integrity. Shapin calls statements of this type, in the context where they were first aired, reactionary, in the sense that they were reactions to the established view.[8] This development in the direction of moral equivalence began in the US, at the point when research was becoming waged work and scientists were becoming employed by industry. When this happened, research became a job and researchers became professionals in a new sense. These new professionals had the same needs and were driven by the same motives as other professionals: they wanted a comfortable life, a good salary, rewarding work, and a satisfactory private life. During the post-war period this view of the scientist as a salaried professional became dominant. From being an individual with particular motives prompted by a commitment to the advancement of science, the scientist became somebody just like the rest of us.[9]

The professionalization of scholarship must thus be seen as a double-edged phenomenon. The apostles of professionalization lived and wrote in a period of change, and must be understood as acting and writing using two totally different and parallel definitions of the scientist at one and the same time. Shapin focuses on the sociological aspect of Merton's writings: the fact that he, as a sociologist, annihilated the need for character in the scientist by outlining an institutionalized system of scientific norms. Institutionalized norms replaced the need for moral fibre, much like a good corset removed the need of for strong stomach muscles. The interest of society does not lie in the superior moral quality its scientists, but in ensuring that they follow the code of conduct established by the scientific community, detailing clear guidelines for how science is to be conducted. Shapin concludes by stressing the institutional character of norms: repetition and rules, but also the social context. Shapin points out that Merton's texts are not only expressions of a scientifically founded insight, but also of a professional strategy, in this case the promotion of the formation of sociology of science as a subdiscipline.[10]

Max Weber's campaign for the value neutrality of university teachers can be subjected to a similar analysis. Swedish historian Peter Josephson has elsewhere given Weberian demands for value neutrality a somewhat different interpretation than previous research. Where it has been usual to see Weber's arguments as the expression of an attempt to professionalize an academic faculty methodologically, Josephson argues persuasively that what Weber tried

8 Shapin 2008, 47.
9 Ibid. 47–91.
10 Ibid. 21ff.

to accomplish was a defence of academic freedom by limiting the same. In order to defend the faculties against state intervention and to gain acceptance for the right of universities to employ individuals who had controversial political or religious beliefs, Weber strived to limit the freedom of individual faculties to preach their own ideas *ex cathedra*.[11] It is plausible, though Josephson refrains from this conclusion, to regard this as strategy for external professionalization. In order to keep and perhaps even expand the autonomy of the profession, its members were to abstain from value judgements.

A conclusion that is close at hand is thus that the proponents of scientific professionalism during the inter-war period undertook an intensive propaganda campaign to exclude morals and values from teaching and research, stating both that values *did* not matter and that they *should* not matter. Of course, it could also be concluded that it lay in their interest to do so, as their claim was the foundation of a professional strategy. When we arrive at this point, it seems that we are also saying that the moral equivalence enshrined in the very concept of a professional strategy (being the expression of *interest*) should make us sceptical both of the importance of an internal set of norms as guidelines in science (the existence of which could be doubted as they are very likely tools for creating public legitimacy), and of the claim to value neutrality in terms of religion and ideology (does not moral equivalence mean that the scientific profession is as imbued with ideology, beliefs and prejudice as the general public?).

Still, Weber and Merton were instinctively normative, in the sense that what they did was to actively promote a moral standard in scholarship, explicitly stating not only what constituted good science, but also who was suitable to make decisions regarding scientific affairs, and on what rationale such decisions should be based. Josephson formulates this very aptly: 'The statement that a value-free science is desirable is in itself a value-judgment, and can therefore by definition not be verified or falsified'.[12]

The question thus boils down to whether stated values and norms should be totally discarded as evidence of the existence of norms? Is it not reasonable to conclude that the normative statements of the apostles of scientific professionalism did not have anything to do with their own views on what science was and should be?

The process that is interpreted by Steven Shapin as a growing conviction of the moral equivalence of science in society, and that went on parallel to the

---

11 Peter Josephson, *Den akademiska frihetens gränser: Max Weber, Humboldt-modellen och den värdefria vetenskapen* (Uppsala: Uppsala universitet, 2005), 11f., 187–192.

12 Josephson 2005, 7f.

development of scholarship as waged work, thus happened at exactly the point in time when a number of texts were published, and gained much attention, that posed moral demands on science and described academics as a moral collective. How to understand this contradiction? Is there an alternative to concluding that the attraction of these texts lay in their usefulness as tools for a professional strategy? Is there a way of understanding academic professionalism that does not discard one of these two insights?

The solution lies, as I see it, in a close reading of Max Weber's 'Science as a vocation'. For Weber, the plea for objectivity meant that academics should refrain from voicing political values within their professional domain. This norm surfaced in Merton's work as 'disinterestedness'. Objectivity was, according to Weber, a condition for professionalism, for science as a vocation. But it was a question of a professionalism that neither excluded strong values regarding work nor strong emotional ties between academics and their work. Science as a vocation was described thus by Weber: 'To affirm the value of science is a presupposition for teaching there'.[13] The university teacher 'fulfils the duty of bringing about self-clarification and a sense of responsibility'.[14]

Whence does the emotional power in this statement come from? What is the ultimate source of its moral command? The simple answer is that it is knowledge itself, and academic work. Work fosters insight, and insight gives birth to clarity; clarity gives meaning to work, and meaning and a sense of meaningfulness is what vocation relies upon. To Weber there is no opposition between professionalism (specialization, value neutrality, facticity) and calling (devotion, meaning, morals); rather, they are interrelated. The legacy of professionalism does not primarily lie in techniques to solve society's problems; it lies in professionalism's ability to create meaning and reason out of what in the strictest sense is purely objective. So what is it about the present science policy regime that leads us to reflect on this?

Economists on Scientific Rationality[15]

A central conclusion in the writings of American economist Philip Mirowski is that the ideology that underpins the present science regime has been formed

13 Max Weber, 'Science as a Vocation', in *From Max Weber: Essays in Sociology*, ed. H.H. Gerth & C. Wright Mills (New York: Oxford University Press, 1946), 129–156.

14 Weber 1946, 152.

15 Another version of the following analysis of scientific rationality from the viewpoint of economists has previously been published in Ylva Hasselberg, 'Demand or Discretion?

largely by economists, with the market as information processor as model.[16] What can substituting the market as a processor of information for professional rationality possibly lead to? This we cannot know at present, but what we can do is to scrutinize the ideas that underpin the process.

What do the economists have to say about the production of knowledge? What type of rationality do they propose? The classic analysis was undertaken by Kenneth Arrow in 'Economic Welfare and the Allocation of Resources for Invention'. The central problem for Arrow was how sufficient new knowledge could be produced in society. He analysed the driving forces of knowledge production, and reached the conclusion that science should be seen, and consequently organized, as a public good, as the inherent character of information as a commodity prevents it from being exploited efficiently by any one owner or interested party. 'Perfect competition' will not be able to achieve optimality in resource allocation, as it will not be rational to invest in information when its production will not primarily benefit the investor. He reached the conclusion that science should be financed and organized by the state, and noted 'the importance of non-pecuniary incentives, both on the part of the researchers, and on the part of the private individuals and governments, who have supported research organizations and universities'.[17] The growth of knowledge is dependent on it being kept separate from the market. As can be understood, Arrow trusted the motivating forces of individual scientists to be quite different from those of, say, the CEO of Microsoft or General Motors. A modernized version of this message is given by Partha Dasgupta and Paul Davis, for although they focus on a partly different problem—namely not how knowledge is best produced, but how to achieve 'a maximal flow of economic rent from the existing stock of scientific knowledge by commercially exploiting its potential for technological implementations'—they hold to the same basic assumption: 'the institutions and social norms governing the conduct of open science […] are functionally quite well suited to the goal of maximising the long-run growth of the stock of scientific knowledge'.[18] The authors, adopting a fairly naive version of the linear model, assume that the diffusion of knowledge (in other

The Market Model Applied to Science and its Core Values and Institutions', *Ethics in Science and Environmental Politics*, 12 (2012).

16 Philip Mirowski, *Science-mart: Privatizing American Science* (Cambridge, Mass.: CUP, 2011).

17 Kenneth Arrow, 'Economic Welfare and the Allocation of Resources for Invention', in *The Rate and Direction of Inventive Acitivity: Economic and Social Factors* (Universities National Bureau, UMI, 1962), 20.

18 Partha Dasgupta & Paul A. David, 'Toward a New Economics of Science', *Research Policy*, 23 (1994), at 518.

words, commercialization) can be more efficient, but they still warn against 'destabilizing and potentially damaging experiments' with the institutional machinery of science. The view of the economists seems to stress the idealism of open science, revealing a rock-steady belief in the ability of its core institutions to withstand any demands for change from society. This is interesting because a comparison with Shapin's modern scientist results in the conclusion that Shapin's scientist is more self-interested and driven by economic motives than the ideal-typical scientist of the economists.

A detailed study of the writings of economists on scientific rationality, however, reveals some peculiarities of this particular version of Mertonian rationality. Arthur Diamond explicitly refers to Merton as the starting-point of his analysis of 'science as a rational enterprise'.[19] So does Paula Stephan in her attempt to formulate the basis of an economics of science.[20] The Mertonian ideal-typical scientist, striving for recognition and trying to establish priority in order to achieve it, is saddled by Diamond with a particular type of utility maximization, which relies on a search for the most elegant theory or the theory with widest scope: in so doing, the scientist maximizes utility, and can thus be considered rational. In order to make this claim, Diamond assumes that 'there must be objective criteria for judging theories'.[21]

Diamond talks about 'stylized facts', which seems to refer to facts that are not located in a historical setting.[22] The social setting, as Wade Hands has so aptly formulated it, amounts to 'summing the individual agents', nothing more.[23] The resulting model of rationality fuses a distorted Mertonian concept of science—value-driven, but stripped of its historical setting and all social and subjective aspects—with an idea of science as an economic enterprise, a specific type of production.[24] It is seemingly easy to relocate Mertonian rationality to a market setting, using a totally different set of analytic concepts. Priority becomes a question of property rights. Production of science becomes a question of the number of publications—an output in terms of countables.[25]

19 Arthur M. Diamond, 'Science as a Rational Enterprise', *Theory and Decision,* 24 (1988), 147–167.

20 Paula E. Stephan, 'The Economics of Science', *Journal of Economic Literature*, 34 (1996).

21 Diamond 1988, 156.

22 Ibid. 152.

23 Wade Hands, 'Caveat Emptor: Economics and Contemporary Philosophy of Science', *Philosophy of Science*, 64 (1996), 107–116.

24 Stephan 1996, 1229.

25 Arthur M. Diamond, 'The Economics of Science', *Knowledge, Technology and Policy,* 9 (1996).

Seeing science as an economic enterprise then allows for questions such as how productivity can be raised; how the reward structures of science can be perfected in order to encourage higher productivity; and how the element of risk involved in the search for new knowledge can be handled so as to remove any impediments to the production of new knowledge. Overall, these questions speak to the larger issue of efficiency. If we assume that scientists are rational agents, maximizing utility in their search for the objectively most elegant theory and then producing papers for the market of ideas, how can we increase efficiency in this enterprise? A question such as this inevitably leads to the problem of reward structures, not least financial rewards.[26] And when the issue of how scientists should be rewarded for their efficient production of new knowledge is raised, the peculiar thing is that the fact that science can be improved through the introduction of an efficient reward structure (implying that scientists are drawn to the greatest financial rewards, and that they compete for them on a market for scientific labour) is not assumed to affect the inner logic of scholarship. The Mertonian inner core of scientific work is left untouched by the assumption that scientists are, at the same time, rational market actors. I would not hesitate to call this version of scientific rationality self-contradictory. The Mertonian character of the economists' perception of science can in fact be seen as a loop, the direction of influence being less than clear. The history and sociology of science contains some very influential borrowings from the field of economics, although, it must be stressed, no influences have come directly from the neoclassical model, which is too far removed from issues regarding individual action to impinge on the sociology of science. Instead, the borrowings have been indirect. When Robert Merton in the 1930s started to think in terms of what forces had given rise to modern science, he fell back on Puritan values as an important explanation. What made scientists different from other people was that they were ready to renounce immediate rewards in order to reach a more distant goal: the long-term accumulation of knowledge. Frugality, self-control, and the ability to postpone or totally relinquish immediate material gain were the hallmarks of the scientist.[27] Merton in his thesis attributed to scientists the same characteristics as the ideal-typical capitalists analysed by Max Weber in *The Protestant Ethic and*

26 Robin Hanson, 'Could Gambling Save Science? Encouraging an Honest Consensus', *Social Epistemology: A Journal of Knowledge, Culture and Policy*, 9 (1995); Diamond 1988; Stephan 1996; Paula E. Stephan & Steven S. Everhart 1998, 'The Changing Rewards to Science: The Case of Biotechnology', *Small Business Economics*, 10 (1998).

27 Robert Merton, *Science, Technology and Society in Seventeenth-century England* (New York: Harper & Row, 1935).

the Sprit of Capitalism.[28] Rationalism is an important ingredient in both concepts, the ability to act according to a means–end rationality being crucial to successful functioning as a scientist or a capitalist. Weber's idea of the birth of the rational capitalist in turn became the single most important historical foundation for the empirical existence of *economic man*. Weber furnished the neoclassical model with a sociological foundation for its most fundamental premises. The theory of the Protestant ethic giving birth to the spirit of capitalism also deflected interest from the question (admittedly not very frequently asked) of whether there really was such a thing as a spirit of capitalism to the question of how its (assumed) existence could be explained. There is thus an evident link between the concepts of the rational economic actor and the rational scientist.

This begs the question of what becomes of the emotional and moral ties to knowledge if we replace CUDOS or the Weberian idea of vocation with the economist's version of Mertonianism described above? This is a problem that intrigues me.

Two Reflections

Sociologically speaking, university teaching staff belong to the category of professionals, and their work is motivated as well as restricted by professional norms of what good science is, and what first-rate work is in relation to a particular notion of good science. The general idea of value-based professionalism has been constantly attacked for the last four decades. Firstly, it is an idea that is politically controversial. It repels socialists and environmentalists, with their critique of established elite groups and scepticism of the idea of a body of knowledge that is not democratically distributed; it repels neo-liberals, with their belief in economic rationality as the basis of all decision-making outside the private sphere, for value-based professionalism and rational choice are not comparable. Secondly, value-based professionalism is also controversial for proponents of the new professionalism—that is, those who advocate a professionalism bereft of its value-base, instead basing it primarily on rules, routines, and technical know-how. The new professionals are competitors of the old. I have my suspicions that all three categories have political ambitions towards claims of knowledge and the power over knowledge and truth claims. The criticism of old-style professionalism for being opaque and self-interested

28 Max Weber, *The Protestant Ethic and the Spirit of Capitalism*, trans. Talcott Parsons (Mineola, NY: Dover, 2003).

(not responsive to society and not really selfless), and the attempts to replace professional judgement with rules, routines, and quantitative measures, are as political and as imbued with subjectivity and self-interest as the claims to truth and honour of the old professionalism. And the new professionalism seems to be remarkably compatible with the idea that scientists are driven by self-interest and are rational economic agents. The work of the technician is for sale and who pays also calls the shots.

Why is this important, given present developments in science, characterized by Philip Mirowski as the rise of the 'globalized privatization regime'? Here I want to make two observations. First, disconnecting science from professionalism comes at a price, since it deprives science of its emotional link to the factual without replacing it with anything else. My hypothesis is that the issue of meaning is fundamentally tied to the driving forces of scholarship. If science is bereft of meaning, we have to ask ourselves how this affects the rationality of scientists and scholars. If the scientists' goal is no longer to achieve a standard of work related to *Sachlichkeit*, what might their goal then be? The obvious answer is the Mertonian norm of originality, so central in modern science. On this point, as in so many instances, the root to this norm lies in the Humboldtian legacy. Originality exemplifies the impulse to do something, to discover something that has not been done or discovered before. This is of course a fundamental aspect of modern science; however, is it emotionally fulfilling to the same extent as *Sachlichkeit*? How does one know when the goal is fulfilled? The answer is that it can never be fulfilled. The originality criterion of science is bottomless, in the same sense as profit maximization as the ultimate driving force behind economic rationality is bottomless. There can always be a bigger profit. And there can always be a new discovery. The goal constantly recedes; it cannot be reached. *Sachlichkeit*, on the other hand, is linked to the core values of modernity. The reproduction of *Sachlichkeit* is fundamental to the meaning of scholarship as well as to the reproduction of society as a modern project.

Second, the normative and emotional aspects of professionalism contribute to a problematization of the simplified Mertonian world-view that is fundamental to the present science regime. Exposing science to the market mechanism cannot be a problem if science itself, as an activity, as a type of work, is immune to market mechanisms. If scientists are driven by a thirst for truth and new findings, we need not worry about their exposure to other motives, and we need not wonder what those motives may do to their questions, theories, and results. The economists' dual scientific rationality (half driven by values, half driven by economic rationality) is at the heart of such simplified Mertonian idyllism, according to which scientists are compartmentalized creatures who leave economic motives behind when they turn on their computers to work. Here, I want to argue that professionalism obliterates the foundation of this claim. Professionals are characterized not by compartmentalization, but by a seamless integration of work and life. The feeling of safety and comfort created by superficial Mertonianism is a false feeling; a feeling that ignores the very essence of professionalism. Taking professionalism seriously, we have a stronger need than ever for normative foundations and the creation of meaning in our work. If we let go of the powerful emotive force that lies in the meaning of quality in our work, we undermine modernity, or what remains of it.

It is true that STS shed its Mertonian heritage decades ago. Doing this, it became something of a taboo in science and technology studies to address the question of norms in science, and to postulate that norms are in some way related to science in a way that makes it possible to distinguish scholarship from other activities through an assessment of the normative structure. According to the 'received interpretation' of science and technology studies, science is what it is, and it is not meaningful or desirable to try to distinguish science from non-science on the basis of its norms. I would beg to differ. CUDOS is today used as a tool for branding and marketing practices that have very little to do with a professionalism based on an emotional attachment to *Sachlichkeit*. It is also used as a tool for the exploitation of an academic proletariat, 'performers' of the real work in laboratories and lecture-rooms. Norms are powerful tools that create good workers if used as motivators. Therefore, the position of science and technology studies has become untenable. It is now time to reclaim the normative structure of science, and to start discussing what it really means—and this is not least a question of enlightenment. If the university as an institution is driven by market forces and profit to such an extent that it affects the production of knowledge and the reproduction of professional judgement, the general public should not be kept in the dark.

CHAPTER 11

The Very Idea of Higher Education

Vocation of Man or Vocational Training?

Sharon Rider

> In general, indeed, the wise in all ages have always said the same thing, and the fools, who at all times form the immense majority, have in their way too acted alike, and done just the opposite; and so it will continue. For, as Voltaire says, *we shall leave this world as foolish and as wicked as we found it on our arrival.*[1]

Introduction

A common objection today to references to Humboldt is that they are merely nostalgic: the mass university has de facto been integrated into a political economic system from which it cannot be separated without serious damage, if not outright destruction, to both.[2] Yet a critique of the present need not be a call for a return to some mythical, glorious past. The following reflections are not concerned with Humboldt's writings as such, nor to the uses or abuses of Humboldt's name in struggle for the identity of the university; but I do try to recall the spirit, if not the letter, of Humboldt's thought, and most especially of the ethos that guided it. In the aftermath of the Napoleonic Wars, Humboldt saw a way to reform the university, not only administratively, but in its very conception. His reforms were to make the university something of use to mankind by entrusting its functions to the scientific spirit of its faculty and the diligence of the students. Vocational schools, it was thought, arrested students'

---

1 Arthur Schopenhauer, *The Essays of Arthur Schopenhauer: Volume 7: The Wisdom of Life*, trans. T. Bailey Saunders, Penn State Electronic Classics Series (2005), 29. (first pub. 1850).

2 An earlier version of this paper was published in Swedish as 'Högskolan bolognese: Otidsenliga betraktelser över den högre utbildningens mål och mening' in Torbjörn Friberg & Daniel Ankarloo (eds.), *Den högre utbildningen—ett fält av marknad och politik* (Hedemora: Gidlunds, 2012). I provide a more detailed discussion of the themes of the Enlightenment idea of autonomy and the Bologna Process in 'Higher Heteronomy: Thinking through Modern University Education', in Sharon Rider, Ylva Hasselberg & Alexandra Waluszewski (eds.), *Transformations in Research, Education and the Academic Market: The Breakdown of Scientific Thought* (Dordrecht: Springer, 2012).

 | DOI 10.1163/9789004271944_013

intellectual and moral development. This is an important point, because the pressure toward the 'vocationalization' of the university is seemingly ever-present and nearly irresistible, and has been so for more than a hundred years, as attested by Nietzsche:

> But the present age is, as aforesaid, supposed to be an age, not of whole mature and harmonious personalities, but of labour of the greatest possible common utility. That means, however, that men have to be adjusted to the purposes of the age so as to be ready for employment as soon as possible: they must labour in the factories of the general good before they are mature—indeed so that they shall not become mature—for this would be a luxury which would deprive the 'labour market' of a great deal of the workforce...the words 'factory', 'labour market', 'supply', 'making profitable' and whatever auxiliary verbs egoism now employs, come unbidden to the lips when one wishes to describe the most recent generation of men of learning. Sterling mediocrity grows even more mediocre, science ever more profitable in the economic sense.[3]

Nietzsche describes an ethos that we recognize as belonging very much to our day. In the last section, I will attempt to show how this ethos is institutionalized as official policy in the Bologna Process.

An Education Suited to an Idea of Man

> I don't know what I want, but I know how to get it.[4]

One of the difficulties attending any attempt to delimit and define the aims of higher education is that any concrete notion about its purpose and function is intimately tied to what a given community at a certain point in time takes for granted about many other important questions. And such aims are of necessity also formulated in the idiom of that community, expressing its ways of thinking and speaking, including what it values and what it despises. Ideas about higher education are necessarily related to hopes and fears about the world that the coming generation is going to inherit, ideas about what they will need to know when they take the steering wheel. In short, the question has to

3 Friedrich Nietzsche, 'On the Uses and Disadvantages of History for Life', *Untimely Meditations*, ed. D. Breazeale, trans. R.J. Hollingdale (Cambridge: CUP, 1999), 99 (first pub. 1874).

4 Sex Pistols, 'Anarchy in the UK' (EMI, 1976).

do with how a nation, a tribe, a people, a state, or a group view life; what they value, what they fear, what they believe in, what they worry about, what they reject, what they hope for: the burning issue of the point and purpose of higher education constitutes a crossroads where a number of the most fundamental human concerns meet. The state of a system of higher education, and what is taken for granted when considering various possibilities and alternatives, is a concretization of the most basic beliefs, values, hopes, and desires of a community, insofar as what the community chooses to do in order to educate their youth shows what they take to be of ultimate importance. In contemporary academic terms, one would say that the idea is historically and culturally situated.

Depending upon which values are supposed or cited, what material or social conditions apply, and which context the discussion arises in, the concept of 'higher education' can refer to different phenomena in the same society. When we talk about 'higher education' in Europe today, for example, we may mean a variety of things: different kinds of publicly financed professional and vocational training; the higher-education system as a whole, including its institutions, such as colleges and universities; the set of relevant laws, rules, regulations, ordinances, and directives regarding the higher-education system as such; and so on. Or the concept can refer to the multifarious components and substance of that education.

In debates concerning the aims of higher education and how these are best to be achieved, these different senses of the term are frequently conflated, or at least not held apart. In what follows, I will attend to this last use of the term, the *content* of higher education, which, I argue, ought to be the touchstone when considering even the other issues. But in point of fact, there has been a marked tendency in recent years to do the opposite—that is, to modify the content of higher education to suit or even be subsumed under other considerations, without any idea of what that content is, on the basis of a layman's concerns without insight into or knowledge of the subject. To take just one example, it has become common to adjust course offerings and programmes to meet what are thought to be the demands of the labour market, say, 'what kinds of engineers are presently in demand?', rather than to consider what kind of education would make someone a good engineer even under unknown circumstances, whatever the future may bring. This shift in focus from content, construed as something of intrinsic value to which all other considerations must be adjusted, to external demands as primary and determinative of content is my first and in many respects most important point. It is often claimed that there is no inherent conflict between the idea of *Bildung* and the strictures placed on higher education by the Bologna Process (most significantly, the requirement of 'employability'), yet as a matter of fact, there is. The conflict has to do with the motivating ideals articulated in the respective models. I want to show in what follows that the fundamental precepts of the Bologna Process are inimical to the very concept of science and scholarship understood as the means of fulfilling, in Fichte's words, 'the vocation of man as such'.[5]

Secondly, as a consequence of the first point, I wish to show that there is a political, or rather a moral, dimension to all the definitions of the aims and descriptions of the purposes of higher education. Every statement regarding what higher education can or should achieve says something about what we value, how we view the relationship between the individual and the collective, and what sort of society we are prepared to build and inhabit, and thus also what we want to change. In other words, the question of what higher education is or should be is ultimately the question of which values and ideals we as a matter of fact embrace, even if we do not explicitly refer to these ideals *as* ideals. Any particular conception of the notion of 'education' is actually a specification of an idea, the use of which is essentially normative and evaluative; every application of it is therefore entirely conditioned by those implicit norms and values. What constitutes a 'good education' is a matter of values, just as in the question of what constitutes 'good art' or 'a just society'.

One way to illuminate buried norms and ideals is to reflect on other ways of thinking and acting, such that the very contrast creates a kind of *Verfremdungseffekt*. Thus I will begin with a review of earlier conceptions of the purpose of higher education. The aim is not to show that the ancient Greeks, for instance, represent a better, wiser, or deeper notion than ours. On the contrary; if anything, the conclusion to be drawn would rather be that any such comparison is invariably misguided and misleading. My point is rather that to understand the Athenian conception of what it meant to educate their young is to understand how the Athenians conceived of life in general: it is to understand *who they were*. We cannot simply embrace the Greek idea of education, and even less revive it, without at the same time radically altering our way of thinking, and even our way of life. What we can learn from considering how they conceived of the task of education is not how to implement their ideas in our society, but rather how much their ideas about education were a part of their way of thinking about man and his place in the world, just

5 Johann Gottlieb Fichte, 'Some Lectures Concerning the Scholar's Vocation', in *Fichte: Early Philosophical Writings*, trans. & ed. Daniel Breazeale (Ithaca, NY: Cornell University Press, 1988). First published 1794. According to Fichte, the vocation of the scholar is 'the supreme supervision of the actual progress of the human race in general and the unceasing promotion of this progress'.

as our ideas about education reflect our values, norms, and our way of life as a whole.

Without a deeper discussion about the intimate connection between education and the values and form of life of which it is an integral part, all planning and evaluation is meaningless. It is perhaps a sign of the times that we devote so much energy to the painstaking development of instruments for the planning, execution, and evaluation of educational institutions, as if the quality of these instruments was of great moment, yet at the same time, we tend to forget the fundamental question of what it is that we are planning for—namely, the future. But the question of what kind of future we hope and plan for is not itself a technical question, nor is it merely a question of economics or administration. Nor is it primarily a political question. It is ultimately a philosophical and ethical question. My observations here should therefore not be read as an attempt at writing a history of educational theory, or a conceptual history of education as a philosophical notion, but as an attempt to place our contemporary ideas about higher education in one historical perspective.

An Idea Suited to a Community: *Paideia* and *Polis*

In classical Athens, the term *paideia* referred to the training considered appropriate for the moral, aesthetic, and intellectual development of the children of the aristocracy; that is, what they needed in order to take on the roles and responsibilities and live up to the ideals that their position in the city-state would place on them. Their education consisted of rhetoric, grammar, mathematics, music, philosophy, geography, natural history, and gymnastics. The goal was the achievement of *arete*, commonly translated as 'virtue' or 'excellence', but perhaps best understood as 'superiority' or 'virtuosity', even in ethical or political matters.[6] The idea seemed to be that education is something that forms, even transforms, the individual by making him more capable of fully realizing the promise of what he was born to be, if only potentially.

Among the important capacities to be formed was that of good judgement, be it in aesthetic, ethical, or political questions.[7] Our interest here is in the idea

6 This notion in itself points to the enormous gulf that exists between the Greek idea of the human being and the modern idea. We can hardly imagine what it would mean to call someone a 'moral virtuoso'.

7 Again, striking evidence of the cultural contrast between the Greeks and us. For the former, the distinction between the moral and the aesthetic was not nearly as self-evident and clear-cut as it is for the modern European.

of a cultivated ability. The teacher's task, given this idea, is not to instil in the pupil a pre-determined set of established opinions, but to induce in him the desire to develop and improve his capacity for sound judgement. The goal of education as such was to enhance the process of intellectual and moral maturity, to teach the young adult to think as one who is sovereign, as one who can exercise discretion and self-control, characteristics which were the *sine qua non* for the patrician way of life as master of the house and citizen (or legislator). Another way of putting the same point is to say that the Greeks thought that a free man was one who has not only the right, but also the duty, to exercise sound judgement in matters concerning what we would call 'values'. Nobility, genuine freedom, and the ability to take responsibility for one's actions and opinions were not innate qualities for the Greeks, but traits that had to be inculcated and cultivated. How this was to be done was thus a question of the greatest concern, for it was a specification of the question 'What sort of human being would express the fullest realization of who we are or strive to be, and how does he become who he is?' Aristotle writes:

> And the same principle applies in regard to modes of life and choices of conduct: a man should be capable of engaging in business and war, but still more capable of living in peace and leisure; and he should do what is necessary and useful, but still more should he do what is noble [*tà dè kalà deî mâllon*]. These then are the aims that ought to be kept in view in the education of the citizens both while still children and at the later ages that require education.[8]

Closely related to the idea of *paideia*, and the purpose which it served, the cultivation of *arete*, was the view that the latter was its own reward: the Greek term *eudaimonia*, commonly translated as 'happiness', implied a kind of growth and thriving, a coming into the full bloom of *arete*, conceived as the realization of man's greatest potential—that is, the achievement of the greatest Good. Thus the ultimate point and purpose of *paideia* was just that: the flourishing of the Athenian citizen as the embodiment of the norms and ideals of the *polis*. Both Plato and Aristotle explicitly contrasted this aim with the kind of training that was thought desirable or necessary for those who were not born to exercise freedom, the worker or tradesman (*banausos*). The *banausoi* were thought to be deformed in body and soul through repetitive and unsound movements, physical and spiritual depletion of their energies, and, perhaps most of all, the child's exposure to and preoccupation with base

8 Aristotle, *Politics* (Cambridge, Mass.: Harvard University Press, 1993), vii: 14, 1333b.

things. Among the most ignoble of such concerns, thought to be the most injurious and spiritually debasing, were activities directed at making money, in contrast to pursuits in the service of higher things: truth, virtue, and the *polis*.[9] Socrates makes the following observation concerning the value of knowing how to perform mathematical calculations:

> Moreover it occurs to me, since the study of calculation has been mentioned, what a subtle study it is and in how many ways it is useful to our purpose, if one studies it for the sake of knowledge and not to buy and sell [*kapeleúein*].[10]

Thus Socrates readily admits that mathematics has a practical value, but insists that this value depends ultimately on its being taught and learned for its own sake, as something that is good to know in and for itself. The Greek *gymnasia* and *palaestrae* were designed to maintain and improve a sound mind in a sound body for a life in the service of the realization of the greatest human good, both for the individual and for the community.

The medieval university had a different task. Its primary purpose was to provide professional training in law, medicine, and, most importantly, theology. *Artes liberales* were considered the basic education, the ground upon which higher studies could be built, the idea being that there are certain things that the student simply had to know about the world and man's place in it before he could embark on a programme of study in the professions, with all the responsibility that the offices in question implied. This basic course of study comprised the *trivium* (grammar, logic, and rhetoric) and the *quadrivium* (astronomy, geometry, arithmetic, and music), which were thought to constitute a coherent system of a kind (as opposed to, say, the study of French or Italian, which were considered worth learning, but not essential elements in the system as such). But the professional training that the student of law, medicine, and theology was to receive was of an entirely different order than the kind of training required to become a skilled craftsman or artisan, due to the nature of the judgements that a priest or lawyer would have to exercise. While a master mason would have shown himself to have total mastery his craft, no

---

9 Plato's critique went further. He thought the influence of commerce on a community to be almost always corrupting, leading inevitably to conflict, discord, and strife.

10 Plato, *Plato's Republic* (Indianapolis: Hackett, 1974), vii: 525d. The Greek infinitive, *gnorizein*, translated above as 'knowledge', should be understood here actively in terms of discovery and recognition of something that one *does*, as opposed to 'knowledge' understood statically, as something that one *has*—say, of a set of facts, theories or doctrines.

longer needing direction or instruction by a teacher, the field over which the priest, doctor, or lawyer needed to be able to survey in order to make sound judgements was the entire natural order, including its Creator and its crowning achievement, man. A professional in this sense was not merely someone who had mastered the tools of his trade, but someone who could be relied upon to exercise his office in a judicious manner within the complex cosmic order which man inhabits. The very idea that someone could take on the responsibilities associated with the Church, the law, or the temple of the soul without familiarity with that order would have seemed absurd.[11]

Further, by entering the university, the student joined a new collective. His identity as someone with this or that native language and this or that home country was augmented by a new identity, as part of a community thought to be higher and more constant than whatever temporary economic or political alliances he might enter into. For the knowledge that he was to attain was thought to be of universal value and applicability. *Universitas*, the community of masters and students, formed a new kind of corporation, with roots most immediately in the institutions of the Church, but dating back to the city statutes of the Roman Empire. The medieval university had the task of producing a certain kind of person, one fitted to fill vital functions in the society of which he was part.

With the Enlightenment, we see a renewed emphasis on the skills and abilities thought necessary or desirable for the practice of self-rule—that is, citizenry. Here we see clear parallels with the line of thought running back to Plato and Aristotle, with the decisive difference that in the eighteenth century, this category included a vastly greater proportion of the population. In a sense, the idea of Enlightenment combined the ancients' aristocratic ideals of autonomy and self-actualization with the medieval conception of universal norms and values applying to all members of a community of thought. For Enlightenment thinkers, what was to be cultivated was the innate capacity for self-legislation (at least in the male half of the population), and the existing social and economic order, regarded as neither enduring nor predictable

11 This interpretation of the meaning of the medieval university is strongly influenced by my reading of J. Ortega y Gasset, *The Mission of the University* (London: Routledge & Kegan Paul, 1946), 43. For a more detailed historical account, see William Clarke, *Academic Charisma and the Origins of the Research University* (Chicago: University of Chicago Press, 2006). My contrast between different ideals of education as expressing the form of life of an epoch is in part inspired by Ortega's ideas about the evolution of scientific thinking from Galileo to the present. J. Ortega y Gasset, *History as a System and Other Essays in the History of Philosophy* (New York: W.W. Norton, 1961).

in an age of intellectual as well as political agitation, revolt, and upheaval, could not serve as the source. Thus it was thought that the capacity for self-determination and self-reliance, a potential that was the birthright of all men, should be cultivated as the realization of man's genuine nature and his proper destiny. Rousseau writes:

> Fit man's education to what man really is. Do you not see that if you try to fit him exclusively for one way of life, you make him useless for every other?...You put your trust in the existing social order, and do not take into account the fact that order is subject to inevitable revolutions, and that you can neither foresee nor prevent the revolution that may affect your children.[12]

Inspired by Rousseau, Kant argued in a similar vein in his lectures on education. He saw education (which, for Kant, included the cultivation of both moral and intellectual qualities) as first and foremost directed towards the actualization of the human potential for freedom and self-legislation (autonomy) in the individual, and ultimately in the species. Kant distinguishes between physical and what he calls practical or moral training.

> Practical or moral [versus physical] education is education toward personality, the education of a freely acting being who can support itself and be a member of society, but who can have an inner value for itself.[13]

The aim of education is not to drill the student in a set of skills like training a horse for dressage, nor to impart specific doctrines, but to *enlighten* him: the point is not to teach him what to think, but how to think.[14] Toward the end of *Die Metaphysik der Sitten* (*The Metaphysics of Morals*), in a section on method in teaching ethics, Kant claims that the core of moral education is to make the student aware that 'he himself can think'.[15] Notice that what Kant is saying here is that thinking for oneself does not arise by itself, but is something

12 Jean-Jacques Rousseau *Oeuvres complètes*, iv: *Émile, ou de l'éducation,* (ed.) Charles Wirz (Paris: Éditions Gallimard, 1969), iii. 468. (first pub. 1762).

13 Immanuel Kant, 'Über Pedagogik', Wilhelm Weischedel (ed.) *Werkausgabe XII* (Frankfurt am Main: Suhrkamp, 1977), 712 (first pub. 1803).

14 Ibid. 707.

15 Immanuel Kant, 'Die Metaphysik der Sitten II: Metaphyschiche Anfangsgründe der Tugedlehre', Wilhelm Weisched (ed.), *Werkausgabe VIII* (Frankfurt am Main: Suhrkamp, 1997) § 50, A 165 (first pub. 1797).

that the human being can learn, something he is capable of being taught to do, through education (*Erziehung*); it is through education, and only through education, Kant claimed, that the human being can achieve his humanity, that is, his autonomy. Further, with a carefully considered and well-devised programme of education, humanity can look forward to a 'future happier human species'.[16]

A University Suited to Its Mission

What I have tried to show in this brief sketch of different formulations of the idea of higher education is the family resemblance between ideas emerging out of fundamentally different historical and cultural contexts. The resemblance that I want to point out has two characteristic features: first, in each case, there seems to be a premium placed on the capacity to see the contingencies of the moment and the circumstances in which one finds oneself, including one's own present interests as well as the influence of others, as something upon which one can exert a greater or lesser degree of autonomous thought or action. Second, in each case, such a capacity for greater autonomous thought and action is not seen as arising spontaneously, but rather as something to be achieved, through the deliberate care of the community. The differences between the ideals described seem to have more to do with what kind of talents, skills, and personal traits are deemed estimable and salient incarnations of the capacity for freedom. Thus, for instance, the difference between Plato's view and Rousseau's resides primarily in the types of character traits they consider necessary or valuable in the world as they know it. Despite these differences, it is clear that in both cases *the aim is to form a certain kind of human being*. If the one who is to be educated is to devote his life to something more than slavery or manual labour—that is, to citizenship and a profession—then he must receive an education proper to the attendant duties and responsibilities. In particular, he must develop his capacity for responsible action, autonomous judgement, and conscientious decision-making, in matters both public and private, practical and theoretical. In short, one might say that the kind of human being to be cultivated through higher education is one capable of sound and independent *judgement*.[17]

16 Kant 1977, 700.

17 One might be inclined to think that the current emphasis on 'critical thought' would constitute an example of this ideal; yet, as I will argue, the automatized systems that have been introduced to train critical thought as a general skill display in their conception and

The educational ideal outlined above bears the stamp of another age, of course. To our ears, it can sound starchy, pompous, impractical and unprofitable. Norms and ideals of the type enumerated here cannot survive in a vacuum; they are born of prevailing attitudes, concerns, controversies, and problems that someone is trying to come to grips with. And it seems to me that the problem that Plato and Rousseau, Aristotle and Kant are grappling with is not whether a people, a society, a state, a community, or a regent has a legitimate interest in the form and content of higher education. From the point of view of the philosophers, the question has never been whether or not the state has the right to involve itself in the business of education, but rather *under which conditions and upon what grounds this legitimate right can be exercised.* In the end, the answer must rest on the nature of that interest. Humboldt, for instance, writes:

> The university always stands in a close relationship to practical life and to the needs of the state, since it is always concerned with the practical affair of training the younger generation. Academies concern themselves only with science and sholarship. University teachers are generally integrated with each other simply through the internal culture and the organisational framework of their disciplines. But regarding their proper business of science and sholarship, they communicate with each other only insofar as they are inclined to do so. The academy is in contrast a society constituted for the purpose of subjecting the work of each member to the assessment of all others.[18]

For just this reason, Humboldt argues for the importance of state control of certain university affairs:

> The right of appointment of university teachers must be reserved exclusively to the state; it is certainly not a good arrangement to grant more influence to the faculties than a prudent and fair-minded body of trustees (Curatorium) would allow. Although disagreements and disputes within a university are wholesome and necessary, conflicts which might arise between teachers because of their specialised intellectual interests might unwittingly affect their viewpoints. The conditions of the

construction an instrumentalist interpretation of that goal which is remote from the ideal described here, and even at odds with it.

18 Wilhelm von Humboldt, 'On the Spirit and Organisational Framework of Intellectual Institutions in Berlin', *Minerva* vol. 8 (1970), 248.

> university are too closely bound up with the direct interest of the state to permit any other arrangement.[19]

I understand Humboldt's point here to be that academic questions—issues involving the actual form and content of research, scholarship, and teaching as such—ought to be assessed on academic or scientific grounds. There is, however, an ever-present danger that the necessity for specialization blinds the academic community to the need for pluralism, and thus can lead to homogenization, standardization and the weakening of science. But furthermore, the university is much more than science and scholarship. Its very existence is based on the societal functions that it is to fulfil. And when it comes to these extra-mural functions, academic considerations are not always primary. This distinction, between the legitimate interests of state and community on the one hand and the interests of science as a pursuit on the other, is reminiscent of Kant's famous argument for freedom of speech.[20] Kant maintained that it is perfectly legitimate for the regent to limit the enlightened citizen's right to public expression insofar as the citizen expresses himself in his capacity as civil servant. As a scholar (or enlightened citizen), he is an equal among equals in the republic of ideas; but as a civil servant, say a clergyman or a tax collector, his rights are limited by his civil function, and he must answer to the state and his fellow citizens, and not only to other scholars. The theoretical distinction between strictly academic (scholarly or scientific) considerations and institutional needs also played out in the nineteenth-century battles over *Lehrfreiheit* and *Lehrnfreiheit*. The issue of how, when, and why these freedoms could and should be curbed or not was debated on and off well into the twentieth century. I take this to mean that this was considered an important question.

Since the expansion and democratization of higher education during the second half of the nineteenth century, the practical function of the university (society's legitimate interest in it) has increasingly taken centre stage in discussions concerning higher education. To begin with, the entry and integration of new student groups (the working classes, immigrants, and, somewhat later, even women) posed new problems and raised new concerns. The university was to provide society with a technically skilled labour force, while at the same time providing these new groups with training in the rights and duties of citizenship in a liberal democracy. They were to be freed from the shackles of

---

19 Ibid. 249.

20 Immanuel Kant, 'Answer to the Question: What is Enlightenment?', in James Schmidt (ed.) *What is Enlightenment: Eighteenth-Century Answers and Twentieth-Century Questions* (Berkeley & Los Angeles: University of California Press, 1996) (first pub. 1784).

ignorance and superstition as well as the ethnic, cultural, and kinship loyalties that bound them and deprived them of the opportunity to participate in the democratic process and debate enjoyed by their more privileged schoolmates.[21] What we have here, it would seem, are two distinct aims, which, at least when formulated in this way, seem difficult to weld into a single goal. When this transformation first began, it was not self-evident that someone who had studied at a business school or a technical college was by definition 'educated'. On the contrary, there was a good deal of discussion about the extent to which practical instruction could or should be integrated into the university. John Stuart Mill, for instance, in his inaugural speech as vice-chancellor of St Andrews in 1867, thought the issue needed to be addressed:

> The proper function of a University in national education is tolerably well understood. At least there is a tolerably general agreement about what a University is not. Universities are not intended to teach the knowledge required to fit men for some special mode of gaining their livelihood. The object is not to make skillful lawyers, or physicians or engineers, but capable and cultivated human beings. It is very right that there should be facilities for the study of the professions. It is well that there should be Schools of Law, and of Medicine, and it would well be if there were schools of engineering and the industrial arts. The countries which have such institutions are greatly the better for them; and there is something to be said for having them in the same localities, and under the same general superintendence as the establishments devoted to education properly called. But these things are no part of what every generation owes to the next, as that on which its civilization and worth will principally depend...Men are men before they are lawyers, or physicians, or merchants or manufacturers; and if you make them capable and sensible men, they will make themselves capable and sensible lawyers or physicians. What professionals should carry away with them from a University is not professional knowledge, but that which should direct the use of their professional knowledge, and bring the light of general culture to illuminate the technicalities of a special pursuit.[22]

21 This was an explicit concern, for example, in John Dewey's classic *Democracy and Education.*

22 John Stuart Mill, 'Inaugural Address Delivered to the University of St. Andrews 1867', in J.M. Robson (ed.) *The Collected Works of John Stuart Mill*, xxi: *Essays on Equality, Law, and Education* (London: Routledge & Kegan Paul, 1984), available at <http://oll.libertyfund.org/title/255/21681>, accessed 11 April 2011. For Mill, astronomy, biology, physics, and

The situation today is somewhat different. In a word, it is not the case that 'the proper function of a University in national education is tolerably well understood'. At the very least, one can say that it is not universally agreed that its proper function is first and foremost to cultivate the capacity for autonomous, sound judgement. Nor is it true any longer that there is 'a tolerably general agreement about what a University is not', if we take that agreement to refer to the general acceptance of the claim that 'universities are not intended to teach the knowledge required to fit men for some special mode of gaining their livelihood'. Quite to the contrary, the systems that have been devised by national governments for auditing and assessing the value and effectiveness of higher education are based on the criteria of standardization, mass production, and, above all, employability. Universities themselves also follow rankings in branch journals in which the value of an education is tied to the average income of its graduates ten years after graduation, and adjust their programmes to emulate those at the top of the list. In other words, what Mill took to be the definitive characteristic of a university is not a relevant factor in considering the value of a course of study in today's policy debates and discussions. To be sure, most policy documents, including those relating to the Bologna Process, stress the role of the university in promoting tolerance, equality, and critical thought. But it is difficult, if not impossible, to demonstrate convincingly the value and efficacy of studying classical subjects such as philosophy, astronomy, and Latin in promoting these goals. At the same time, it goes without saying that such subjects are unlikely to achieve the same success in attracting students or guaranteeing their future employment as more practical programmes. The only argument for their continued existence would have to be Socrates': however useful they may turn out to be, they will only be valuable if they are taught and learned as worth knowing in themselves. But in a society in which value is strictly measured in terms of foreseeable practical utility and economic growth, it is perfectly natural that 'employability' serves as the umbrella term for the technical competence and vocational skills required by industry, the market, and the public sector. Thus a good education is by definition an education that produces highly skilled workers who are in great demand. An excellent education is one in which the students achieve such a high degree of technical accomplishment that they can not only follow technological developments and their attendant economic benefits, but actively contribute to them. An excellent university is hence one that produces innovations and innovators.

mathematics were as important and even indispensable elements of a general or liberal education as law, political science, and, for reasons that can be understood in terms of 'multicultural awareness', classical Greek.

This ideal differs more radically from previous ideas than any preceding change in terms of its fundamental shift in focus from ensuring that the coming generation consists of 'capable and sensible men' to a concentration on commerce and competition. That this transfiguration is substantial becomes clear if one considers that Plato and his contemporaries saw the latter as sources of community dissolution, albeit one which could be held in check by seeing to it that the free citizens of the community were given a proper education.[23] Where Plato saw the role of education as indispensable for the creation and sustenance of a sense of commonality precisely by training the capacity for autonomy in those who had the task of fulfilling the ideals of the *polis*, current trends in higher education would seem to see education as an efficient way of dissipating the sense of community (national belonging, disciplinary integrity, and so forth) by enforcing a heteronomous framework in which the student is perceived as a token of an amorphous mass of mobile, potential employees, distinguishable primarily by the set of skills he can demonstrate through a standardized certification process.[24]

An Education Suited to the International Market

Up until this point, I have assumed the reader's acquaintance with what I have called 'current trends', and which I have exemplified simply by allusions to the Bologna Process *tout court*. I think it important, however, to specify with a few examples more precisely what I mean, even though it will mean moving from philosophical reflection on the nature and purpose of the cultivation of human thought by some of the greatest minds in Western history to pamphlets and proclamations produced by committees with the explicit aim of structuring, standardizing, and formalizing the European system of higher education. The

23 I have intentionally avoided referring to any of the established theoretical positions regarding the philosophy of education. In my view, the majority of the overtly normative positions ('perennialism', 'essentialism', 'progessivism', and the like) in the main share the view that higher education ought to contribute something more both to the individual and to society than professional or vocational skills, if we are to justify the existence of institutions such as universities. Where there is disagreement, it has to do with what one takes these higher or broader aims to be, and how these are best attained. I have not taken a clearly defined stance here, although I do suggest that the capacity for judgement (a philosophically difficult concept) is central.

24 On the role of education in the creation of the modern European nation-state and the deterioration of both, see Bill Readings, *The University in Ruins* (Cambridge, Mass.: Harvard University Press, 1996).

documents are every bit as officious and arid as one might expect, but I take it as one of my tasks to sift through the layers of technical terminology in order to illuminate the conceptual implications embedded in them.

I have selected as my illustration a report written by the Bologna Working Group on Qualifications Frameworks.[25] The expressed aim of the report is to make recommendations and 'offer advice on good practice' in the elaboration of national frameworks for higher-education qualifications. Now, this is merely one in a slew of regional, national, and European reports, commissions, communiqués, statements, schemes, and manifestos to have emanated from the Bologna Process since its inception, but it is representative of the style and content of the majority of them. There is nothing in this particular report that conflicts with the basic ideas behind 'Bolognization'; on the contrary, I have chosen it because it is genre-typical.

Due to a lack of established consensus on terminology, the authors offer a glossary of terms used in the report. Among these is the crucial notion of 'learning outcomes', the axis on which all educational activities are supposed to turn. 'Learning outcomes' are defined here as 'statements of what a learner is expected to know, understand and/or be able to do at the end of a period of learning'. Consequently, a 'credit' is to be understood as 'a quantified means of expressing the volume of learning based on the achievement of learning outcomes and their associated workloads'. Learning outcomes are pivotal because they 'represent one of the essential building blocks for transparency' with regard to qualifications. With 'outcome-based approaches', they have 'implications for qualifications, curriculum design, teaching, learning and assessment, as well as quality assurance' and 'are thus likely to form an important part of 21st century approaches to higher education (and, indeed, to education and training generally) and the reconsideration of such vital questions as to what, whom, how, where and when we teach and assess'. Because of this profound influence, learning outcomes are 'important tools in clarifying the results of learning for the student, citizen, employer and educator'.[26]

One might wonder how it is that a formal statement of what the student is expected to know at the end of a course of study can have such a fundamental effect on higher education—on the 'what, whom, how, where and when' of university teaching. The answer is that the perfunctory definition with which the report begins actually 'represents a change in emphasis from "teaching" to

25 Bologna Working Group on Qualifications Frameworks, *A Framework for Qualifications of the European Higher Education Area* (Copenhagen: Ministry of Science, Technology, and Innovation, 2005).

26 Ibid. 38.

"learning" typified by what is known as the adoption of a student-centred approach', which 'produces a focus on the teaching—learning—assessment relationships and the fundamental links between the design, delivery, assessment and measurement of learning'. And this, in turn, is important because it 'implies that the manner of the achievement of a qualification is not as important as the achievement of the qualification itself'. Finally, this is shown to be desirable because, among other things, it facilitates employment, since prior work experience can now be fully accredited without the process of formal academic training.

But what does it actually mean to implement 'learning outcomes' in practice? On the face of it, not much. Instead of syllabi, which are formulated in terms of the aims and purposes embedded in the 'course content'—what the lecturer as a practitioner of her discipline considers central—the course plan is now determined on the basis of what the student is expected minimally 'to know, understand or do' in completing the course successfully. The difference is merely that now the student (as well as future potential employers) will be in possession of a specific formulation of the minimal requirements, and the lecturer must base her teaching, along with her examinations and assessments, on those specifications.

In this respect, course plans in the Bologna Process resemble little so much as contracts, whereas syllabi are generally neutral descriptions of a subject matter and a plan of action for how it is to be assimilated. This shift in focus from subject matter to expected outcomes means that both teachers and students are to conceive of education in a new way: the old way of thinking 'first—subject matter, then—teaching, then—examination' is replaced with 'first—criteria for assessment and accreditation, then—techniques, methods, and systems for fulfilling the requirements stipulated at stage one, then—expected and achieved outcomes'. The latter is called 'constructive alignment', the implementation of which is thought to ensure a more transparent system of qualifications and progression between levels of education. The basic idea is that having achieved a certain level of qualification, the student and educational institutions can 'check off' the relevant step on the protocol and thus predict whether the student is qualified for the next level.

Let us consider the implications of this almost imperceptible administrative shift in the organization of teaching practice. There is one assumption in particular built into the acceptance of the scheme that merits consideration: the idea that it is both possible and desirable to lay down in advance the expected results of any course of study, regardless of its content, whether or not it is theoretical or practical. Let us consider the former case, in other words what it means to specify what a given student should be able 'to know, understand or do' having studied, say, philosophy for three months. Most students who study philosophy for a term walk away thinking that Plato said that we are all living in a cave, that Kant believed that it was better to reveal the location of the Jewish family hiding in your attic than to lie to the SS, and that Nietzsche argued that everything is about power. Some students develop their capacity for critical thought and sound judgement after a course of study in philosophy, but many do not. That they can learn to apply certain techniques to meet with specified requirements stated at the outset does not mean that they have started thinking critically or philosophically.

Now it may be objected here that it is up to the lecturer to see to it that the criteria are so precise, and the application of them so fastidious, that only a student who has genuinely learned to think critically, at least at a rudimentary level, will be deemed to have fulfilled the course requirements. But no philosopher has ever succeeded, despite innumerable attempts, to state the nature of critical thought once and for all. It seems unlikely that a garden-variety professor at a run-of-the-mill European university should succeed where Kant failed. Thus if the application of such instruments is to have any meaning, university teachers and administrators should concern themselves with formalizing and automating professional judgement. Indeed, the more formalized and mechanized the instruments, the less there is need for judgement at all. And this is explicitly the point, since two of the main goals of the Bologna Process, mobility and transparency, are two sides of the same goal: interchangeability. The idea is that a course of training in, say, internal medicine in Lille should be equivalent in form and content to the 'same' course in Umeå; something that facilitates the exchange of students and staff within the EHEA, and, ultimately, the entire European workforce. The transparency provided by the Bologna Process provides employees with a universally recognized certification, and the employers with a standardized means of comparison and valuation.

Thus the value of 'learning outcomes' lies in the transparency and measurability they create. The aim is to inform the student and her prospective employer what it is that she has attained in her course of study in terms of specific competences and levels of attainment. Her attainments are to be assessed according to a (within the EHEA) universally recognized standard, including those for general competences such as 'written and oral communication', 'information search and retrieval', and so forth. What is assured by these measures is not the quality of the education itself, but of the system of certification.

Philosophically speaking, the basic assumption at work here is that subjectivity is merely the opposite of objectivity as construed in the everyday sense as tantamount to arbitrary biographical or biological determinations or idiosyncrasies. The value of the cultivation of the capacity for and exercise of judgement as a *sine qua non* of science and education is, given this assumption, almost incomprehensible. Judgement is assumed rather to belong to the domain of the aesthetic and perhaps the ethical, but has no place in scientific thinking per se.

Now compare this call for mechanical objectivity with the emphasis on the cultivation of judgement found in past conceptions of the purpose of higher education. In antiquity, the Middle Ages, and the Enlightenment, judgement was inescapable as long as there was evaluation. Judgement is something necessarily exercised by the subject, however one conceptualizes the grounds on which it is exercised. Mechanized quantification does not do away with judgement; it merely weakens, impairs, and hides it. In this respect, formalized systems of verification and assessment are undemocratic, since the norms on which they are based are not transparent. You cannot reason or debate with a thing, and what formalized procedures and protocols do is to reify the norms involved in the activity of assessment and evaluation so that considerations alien to the system are locked out at the outset, thus crippling judgement and disenabling reasoned discourse.

I have argued that higher education has always been linked to the exercise of autonomous judgement. This is most especially and most explicitly the case during the Enlightenment, the period when the modern research university came into being. Kant indeed argues for the need for at least one faculty, the 'lower' philosophical faculty, to serve as the foundation for the other, more worldly faculties (theology, medicine, and law) and their 'business', by virtue of its freedom to submit their assumptions to critique.[27] The philosophical faculty, in contrast to the others, should be fully independent of external authority with regard to science and scholarship (which naturally included university teaching), and this as a safeguard for the continuation of science as such, including the applied sciences of law, medicine, and theology, the work of which was conditioned by external authority. The members of the committee on qualification frameworks for the Bologna Process do not share Kant's view:

> 'Externality' is increasingly recognised as an essential part of quality assurance, and so it should be within the development and application of new national qualifications frameworks. For such frameworks to be of benefit to stakeholders, including intending and current students, and

27 Immanuel Kant, *The Contest of the Faculties*, trans. Mary MacGregor (New York: Abaris Books, 1979) (first pub. 1798).

> their employers, the frameworks need to be expressed in terms that are understandable and relevant.[28]

Aside from 'intending and current students and their employers', who are these 'stakeholders' who are to guide the development and application of qualifications frameworks (and thus, via the formulation of 'learning outcomes', 'vital questions as to what, whom, how, where and when we teach and assess')?

> The stakeholders may include: learners/students; providers of education and training; government and appropriate government agencies; awarding bodies; higher-education professors/teachers; employers and the business sector; trade unions; community and voluntary organisations; professional bodies; etc.[29]

Regardless of whether the external authority is a prince, a politburo or, as one concludes from the quotation above, just about everyone, it is still an external authority. The more 'externality is recognized as an essential component of quality assurance', the further away we have moved from the idea of autonomy and the development of the capacity to exercise judgement. Indeed, a synonym for 'externality' would be, in Kantian terminology, heteronomy, i.e. the subordination of an activity to principles alien to that activity laid down by an external authority, as opposed to autonomy, self-legislation in accordance to internal principles stemming out of the aim and meaning of the activity in question.

Conclusion

I make no claims to have the answer to what higher education is or should be in essence; indeed, I have been at pains here to point out that there is no universal concept, but only different forms of life expressing and perpetuating themselves. Yet one can see what a culture or a society is really about by looking at what it does to prepare its youth for the future, and how it articulates the motives and rationale behind this preparation. There are any number of relatively recent books, many of them quite good, describing today's higher-education policy with a variety of fitting titles: *Academic Capitalism*, *University Inc.*, *Universities in the Marketplace*, *The Triumph of Emptiness*, *Killing Thinking*,

28 *Bologna Working Group on Qualifications Frameworks 2005*, 53.

29 Ibid.

University in Ruins, and so on. And there is much to be said about the instruments of this transformation and its effects; in particular, the loss of a sense of mission. In a society in which human judgement is considered unreliable, and by necessity subjective and even partisan, it is only natural that it will be replaced by formal protocols. In such a society, the development of character is not a possible goal, since character, again, is something judged, not arithmetically measured. The notion of self-government is barely conceivable in a form of life in which there is no government, but only governance; no accountability, but only accountancy. Yet it seems to me that to bemoan the loss of the sense of mission in higher education is simply to lament the fact that we have the form of life we have—that we are who we are.

Index

www.ingramcontent.com/pod-product-compliance
Lightning Source LLC
LaVergne TN
LVHW010657300926
849161LV00005B/1152
9789004271920